Republic F-105 Thunderchief

Operations in Southeast Asia

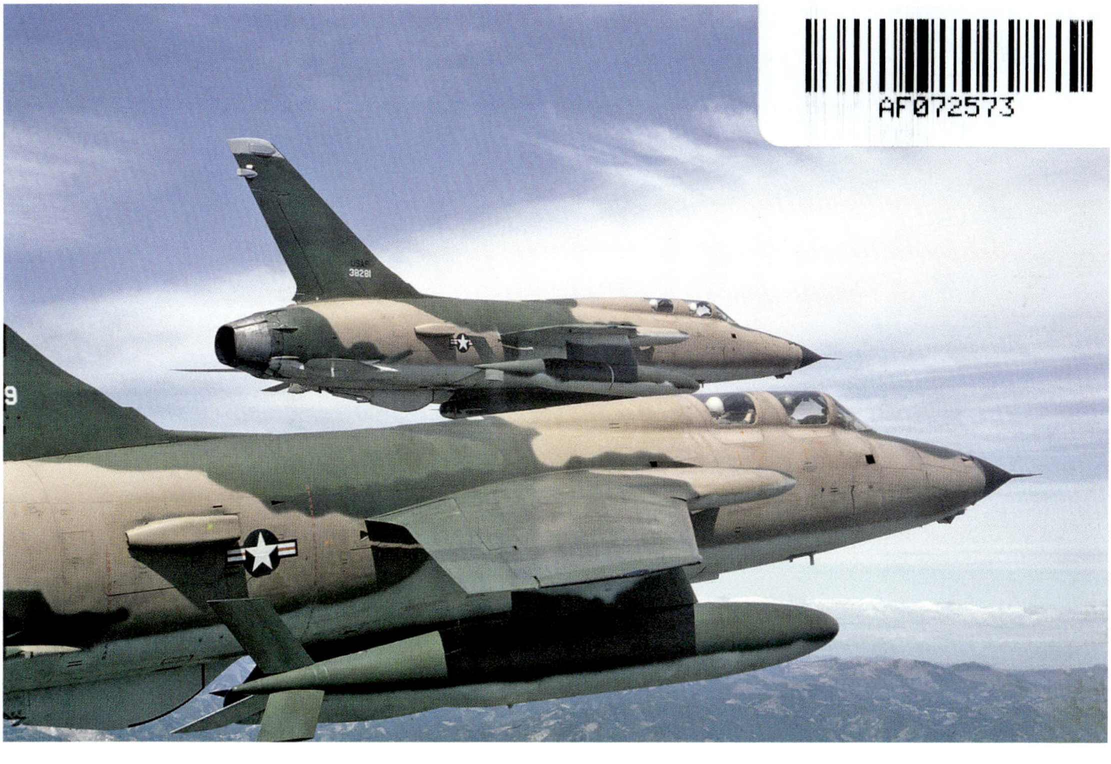

THEO VAN GEFFEN AND
CMSGT GERALD ARRUDA (USAF, RET.)

KEY Books

HISTORIC MILITARY AIRCRAFT SERIES, VOLUME 14

Front cover image: F-105D 60-0522 of 357th Tactical Fighter Squadron (TFS) during prestrike refueling. Two MiG-17 kill markings are shown. Only one was confirmed by the United States Air Force (USAF) – the one by Maj Robert Rilling of the 333rd TFS on May 13, 1967. (USAF, TSgt Thomas Sanchez)

Back cover image: A flight of four 355th TFW F-105Ds behind a KC-135A Stratotanker during prestrike refueling. (USAF, John Evans)

Title page image: In April 1967, the first four "Ryan's Raiders" F-105Fs arrived at Korat Royal Thai Air Force Base (RTAFB) from Yokota AB. The photo shows F-105Fs 63-8269 and -281 en route to Kadena for a pit stop. (Nick Donelson)

Contents page: By December 31, 1972, nearly all F-105Gs of the 17th Wild Weasel Squadron (WWS) at Korat had received shark mouths. (Coll/TvG)

Authors' Note

The history of the F-105 Thunderchief and the units it was assigned to is a rather complicated and extensive one. No book has yet been published that tells the complete story. This book, the previously published *Republic F-105 Thunderchief: Peacetime Operations*, and future third in the trilogy, which will deal with modifications and resulting new missions, will not either. We, after discussions with the publisher, decided to focus on a couple of specific subjects, so that they at least come as close to an almost complete coverage as possible. The photos will tell the rest.

Acknowledgements

Special thanks go to Sandor Kocsis and Howard Plunkett. Thanks to the Air Force History Support Division (Pat Engel and Richard Wolf), the Air Force Historical Research Agency (Archie Difante, Tammy Horton, Sly Jackson and Sam Shearin), and the Cradle of Aviation Museum (Joshua Stoff).

We are indebted to the photo contributors as credited in the captions, plus the following persons:

Dick Baughn, Marty Case, George Cully, Dag Damewood, Lucky Ekman, Dave Everson, Dave Graben, Dave Ferguson, Dave Groark, Frank Hayes, Bob Krone, Tom Lockhart, Al Logan, Donovan McCance, Hans Ragay, Bill Ramage, Jack Redmond, Henk Scharringa, Ed Skowron, Billy Sparks, Larry Van Pelt, Vic Vizcarra, Wayne Waddell and André Wilderdijk.

Sources

Histories of the 4th, 18th, 23rd, 355th, 388th, and 6441st Tactical Fighter Wings (TFW), 41st Air Division (AD), 5th and 7th Air Forces (AF), Tactical Air Command (TAC), Pacific Air Forces (PACAF), Sacramento Air Materiel Area (SMAMA), and Tactical Air Warfare Center.

"Comparative analysis of USAF fixed-wing aircraft losses in Southeast Asian combat," AFFDL-TR-77-115, Dec 77; F-105D/F/G Flight Manual, Change #7, May 15, 1978; *Gradual Failure* by Jacob Van Staaveren; "Electronic Countermeasures (ECM) in the Air War against North Vietnam" (K168.01-52); "In-Country and Out-Country Strike Operations, Volume IV- Support, Research and Development" (K717.0421); "USAF Buildup in Thailand 1965 and 1966" (K750.04-9); "Analysis of Air Operations SEA" (K717.308-1); "Barrel Roll 7" (K717.0414-4); and "USAF Operations from Thailand 1954-65" (K717.0423-29).

Published by Key Books
An imprint of Key Publishing Ltd
PO Box 100
Stamford
Lincs PE9 1XQ

www.keypublishing.com

The right of Theo van Geffen and Gerald Arruda to be identified as the authors of this book has been asserted in accordance with the Copyright, Designs and Patents Act 1988 Sections 77 and 78.

Copyright © Theo van Geffen and Gerald Arruda, 2022

ISBN 978 1 80282 243 4

All rights reserved. Reproduction in whole or in part in any form whatsoever or by any means is strictly prohibited without the prior permission of the Publisher.

Typeset by SJmagic DESIGN SERVICES, India.

Contents

Chapter 1 USAF Reaction to the Gulf of Tonkin Incident 4

Chapter 2 Temporary Duty at Korat, Da Nang, Takhli and Tan Son Nhut 15

Chapter 3 First Blood 39

Chapter 4 Permanent Units at Korat Royal Thai Air Force Base 46

Chapter 5 Permanent Units at Takhli Royal Thai Air Force Base 62

Chapter 6 The 355th TFW and July 20, 1966 72

Chapter 7 Southeast Asia Farewell to the F-105 Thunderchief 80

Epilogue 90

Chapter 1
USAF Reaction to the Gulf of Tonkin Incident

On July 31, 1964, the USAF had some 155 aircraft in SEA, including six F-100Ds, four (T)F-102As and four KB-50Js in Thailand, and 50 C-123Bs, 15 A-1Es, and ten RF-101Cs in South Vietnam.

Operation *Pierce Arrow*

After the alleged August 2 and 4, 1964, attacks in the Gulf of Tonkin by North Vietnamese patrol and torpedo boats against the destroyers USS *Maddox* and *C. Turner Joy*, the US reaction included a retaliatory attack, Operation *Pierce Arrow*, in which A-1 Skyraiders, A-4 Skyhawks and F-8 Crusaders from the carriers USS *Ticonderoga* and *Constellation* were launched. In 64 strike sorties, they struck the Vinh Petroleum Products storage and naval bases at four different locations. Two aircraft were lost, an A-1H and an A-4C. The pilots were killed and taken as prisoner of war (POW), respectively. Two aircraft were damaged, but the pilots were able to return to their carrier.

In addition to *Pierce Arrow*, Secretary of Defense (SECDEF) Robert McNamara announced, with President Lyndon Johnson's approval, the deployment of additional US reinforcements: from PACAF assets, 36 B-57Bs, 12 (T)F-102As, two RF-101Cs, four F-100Ds, four KB-50Js, and 18 F-105Ds, going to SEA; from TAC assets, as part of a Composite Air Strike Force (CASF) *Pierce Arrow* "64", two F-100D/F squadrons with 36 aircraft (to Clark Air Base [AB], Philippines), an F-105D squadron with 18 aircraft (to Yokota AB, Japan), a reconnaissance element of six RF-101Cs and a Photo Processing Cell (to Kadena AB, Okinawa) and an airlift element of 44 C-130B/Es (to Clark AB and Naha AB, Okinawa). For deployment of the TAC fighters, Strategic Air Command (SAC) provided 48 KC-135A Stratotankers, operating from Hickam AFB (Hawaii) and Andersen AFB (Guam). These flew 172 sorties.

The first aircraft to deploy were four F-102As and two TF-102As of the 509th Fighter Interceptor Squadron (FIS) on August 5 from Clark to Da Nang AB in northern South Vietnam. The first aircraft launched at 02:30L. At 08:00L, all were on alert, two at five-minute, two at 15-minute and two at one-hour. The alert posture was retained throughout 1964. The Detachment (Det) was known as 509th ADVON 2.

36th TFS

On August 5, the Joint Chiefs of Staff (JCS) alerted certain forces under selected portions of CINCPAC (Commander in Chief, Pacific Forces) Operations Plan (OPlan) 37-64. Included was the alert order for one PACAF F-105 squadron to deploy from Yokota to Korat. Also, one Strike Command F-105 squadron in the US was alerted to replace the deployed PACAF unit and assume its Single Integrated Operational Plan (SIOP) commitment at Osan, Republic of Korea.

At 09:30Z(ulu), August 5, CINCPAC directed PACAF to deploy the F-105 squadron to Korat under PACAF Operations Order (OPORD) 145-65, after Thai clearance was received from US Ambassador Graham Martin in Bangkok. Also, KB-50Js of the 421st Air Refueling Squadron were to deploy from Yokota to Kadena to refuel the Thunderchiefs on their route from Yokota to Clark. Four KB-50Js would then deploy to Takhli Royal Thai Air Force Base (RTAFB). Three hours later (12:30Z), PACAF informed 5 and 13AF.

In the meantime, 5AF frag order #1 to 5AF OPlan 37-65 had designated the 36th TFS (41AD) as the deploying F-105 squadron. Lt Col Don McCance, the squadron commander, then called a meeting and informed everybody of the imminent deployment. Special Orders (SO) T-1416 and -1418, issued on August 5 by the 441st Combat Support Group (CSG) at Yokota, directed 28 officers, including 26 pilots, of which two were spares, and 58 maintenance/support personnel to proceed on/or about August 6 to Clark on Temporary Duty (TDY) for approximately 179 days, for the purpose of performing an operational mission in support of 5AF OPlan 37-65. All pilots, except one, were assigned to the 36th TFS.

On August 5, Dave Hubbard, the maintenance officer, led an advance party to Korat, while crew chiefs were sent ahead to Clark to turn around the F-105s after landing there. The next day, 20 F-105s, including two spares, departed non-stop to Clark, with the first 14 at 09:00L and the final six at 13:30L. The aircraft were refueled by nine KB-50Js just past Kadena. The tankers were four hours on station and offloaded 253,000lb of fuel. After a flight of almost four hours, the F-105s landed at Clark, where they were held pending diplomatic clearance and the passing of a typhoon. At 14:56Z, 5AF informed 41AD (Air Division) that clearance was confirmed and directed them to execute the deployment to Korat as soon as possible. Although all aircraft were ready at 07:00L on August 7 for departure, weather and the lack of TACAN (Tactical Air Navigation) at Korat delayed the departure. The only approach available was Automatic Direction Finding (ADF), but the F-105 lacked this system. This meant a mobile TACAN had to be flown in. Finally, on August 9, the first flight of four aircraft launched at 08:30L, landing at Korat some two-and-a-half hours later. The next flights followed at 15-minute intervals, with the final of 18 aircraft landing at 12:27L. As soon as they had landed, they were turned around. At 15:00L, eight aircraft were placed on one-hour air defense alert, configured with the 20mm Vulcan gun only. The 2AD commander, Maj Gen Joe Moore, visited and briefed the Squadron about its mission. The next day, all pilots flew a local orientation sortie. While at Korat, the unit was attached to 2AD (Tan Son Nhut AB) for operational control.

The two spares were flown by Capts Dave Everson (80th TFS) and Bob Becker (35th TFS). Capt Everson recalled:

When we were taking off from Yokota, at least a dozen television vans with cameras mounted on the roof were filming our departure. Many Japanese civilians worked on base and it was no secret we were going. We entered the typhoon just past the southern tip of Taiwan and were in it until we broke out about a mile out on final to Clark. I estimated my head was three feet from my lead's wingtip light and I still lost sight of it several times. The refueling by KB-50Js was the last time I was to see a KB-50 and did not feel sorry. Refueling off a KB-50 was hard work. It trailed a forty-foot hose with a funnel shaped basket at the end. The hose was flopping around all the time. The F-105 was easier than the F-100 Super Sabre because the 105 refueling probe was forward of the cockpit so we could steer it more easily. As we were left at Clark with two broken aircraft, we were put to work figuring F-105 mission profiles for a variety of missions and weapons loads, M-117 bombs and rockets, no missiles. Those profiles did not exist, at least not in the Pacific theater. They showed how far we could carry specific bomb loads and get back, with a reasonable fuel reserve for combat time. We provided them to 2AD, which presumably sent them on to Korat. This was mostly so the people who publish frag orders had an idea of what they could realistically expect F-105s to do. All were planned without tanker support, as the planners did not have any idea what the war was going to become. They also had no idea of the F-105 capabilities. When that was done and the aircraft repaired, we returned to Yokota through a refueling stop at Kadena.

Korat
Korat's caretaker staff consisted of one officer and 14 airmen. They were given less than eight hours to prepare for the arrival of ultimately some 500 personnel and 18 F-105s. The base could be considered, at best, a bare strip with almost no facilities. It lacked a control tower, and a temporary one had to be set up. Assistance was requested and received from the US Army's 9th Logistical Command's

Detachment at Camp Friendship, located close to Korat: 300 personnel were quartered in the USAF area and 200 at Camp Friendship. The Army loaned beds and bedding and all subsistence was supplied for the first two weeks. Four days after the Squadron's arrival, the Army constructed more quarters, added to the dining hall, while latrine and shower facilities were completed. Lodging was not the only problem, however. Very little equipment was on hand to offload gear, there was no space for storage, munitions were offloaded by use of a forklift onto Army trucks and a flatbed trailer and initially stored on the parking ramp and then moved to a hastily prepared munitions storage area close to the flight line. However, munitions-handling personnel and equipment did not arrive until a week later. Maintenance was carried out in the weather, aircraft ground equipment was insufficient, and spares and bench stocks were not provided in adequate quantities to sustain training and combat sorties. The electric power supply went off for as long as a day at a time, with continuing loss of communications.

After Det 10 of the Eastern Air Rescue Center at Maxwell AFB (Alabama) had been alerted on August 6 for deployment to SEA, its two HH-43Bs and personnel were airlifted to Korat by C-124C Globemasters. The unit arrived on August 13–14 and became Det 4, Provisional. Its mission was local base rescue. The Huskies were reassembled, and the first sortie flown on the 15th. Additionally, the Ground Control Approach (GCA) became operational.

Search and Rescue

Although the Korat F-105s were better suited to accompany the regular Laos Operation *Yankee Team* RF-101C Voodoo escort missions, they were flown by Takhli F-100D Super Sabres. The reason for this was a lack of Thai approval. Four Thunderchiefs were on five-minute cockpit alert to fly SAR (Search and Rescue)-supporting missions over Laos in case of a downed aircraft (this changed after September 21, 1964, when the 35th TFS deployed to Korat, its pilots not only flying SAR-supporting but also RF-101 escort missions). The aircraft were configured with two LAU-3/A pods with 19 2.75-inch rockets each and 1,000 rounds of 20mm high-explosive incendiary (HEI) ammunition. The first such mission was flown on August 14, when four F-105s were called upon to provide air cover for the rescue of a downed aircrew in the Plain of Jars in Laos. Number two in the flight was 1Lt David Graben. He stated:

> Our call sign was probably Kilo as we were Kilo Flight. Capt Jack Stressing was #1, I was #2, Capts Rod Beckett #3 and Marty Case #4. We sat alert in the cockpit, ready to go. There was quite a bit of competition for the "privilege" of sitting in a hot cockpit for an hour or two. I usually read a book, but it was hot and unpleasant. The crew chief was at the aircraft, but sometimes in a nearby truck with the air on. It was not a pre-planned mission. We received a frag order from 2AD in Saigon after the US ambassador to Laos had requested assistance in the rescue effort of a civilian Air America T-28 pilot who went down and [was] eventually rescued successfully. A gun site was slowing down the effort. An Air America Caribou pilot was our FAC, Forward Air Controller, and rescue coordinator. No other aircraft were involved as far as I know, although some T-28s were in the area. We briefed at the aircraft, had everything we needed, although we knew nothing about area defenses. We wore a "vest" with a lot of goodies, including a .38 revolver and flares. And yes, I was pumped up for the mission, as that's what you trained for.

Lt Graben continued:

> We took off at about 10:00L. As there were no KB-50Js in the frag order, our aircraft were configured with one centerline 650-gallon and two 450-gallon fuel tanks, two LAU-3/A pods on the outboard pylons and the guns were "hot." After arriving in the target area, I rolled in on the target after Jack, but

did not fire the gun or drop any ordnance. I was out of range when I came under fire. The red-orange tracers were floating towards me from directly at 12 o'clock. As soon as I felt the "thump," I knew I was hit. I was about 1,000 feet above ground level and broke to the right. It took a few seconds before the fire light came on, letting me know for sure I was hit, although I did not know where. I called, "Two's off, I've been hit." Jack reacted with, "Get rid of your stores and head south." I lit the burner, jettisoned all stores and started a climbing left turn. Jack joined up when I levelled off at 31,000 feet. He looked me over and told me, "Davy, you have a rather large hole in your left stab and it looks like the bottom of the tail ahead of the speed brake is burned through." The left stab moves with the right. It was functional. The fire had gone out by now, but I experienced a utility hydraulic failure at about the same time. I did think about ejecting, but to get as far away as possible from the bad guys, I lit the burner again. When I crossed over into Thailand, things started to look brighter and I planned the ejection to be my final option. I ejected from an F-100F on December 7, 1962, and was not afraid to do it again. There was not much of a pitch problem until I slowed down on final when it seemed to take a bit more effort to control it. It was never a real problem, except for the loss of normal braking (emergency braking was used), emergency gear (there was an emergency system to lower the landing gear) and flap extension. I did not think the landing would be too hairy. Jack did chase me down on final and let me land first. Korat was prepared with the CRS [Crash and Rescue Services], being ready. After touchdown I deployed the hook, but did not need it. After a full-stop I shut down on the runway as nose steering was inoperative. The sortie had taken two hours and fifteen minutes. After I got out, I looked the aircraft over and realized the damage was greater than I thought. Also, that I was lucky to get back on the ground. I was not wounded. The aircraft was not on the runway very long, enabling the other three F-105s to land. The Squadron put me in for a Distinguished Flying Cross, but the USAF reduced it to an Air Medal: we were not at war! The citation did not go into any specifics.

Classified

The one 35th TFS pilot to receive orders for TDY was Marty Case, who was on the brink of getting married. Marty recollected:

Only one 35th pilot was selected to join the 36th TFS. Some of the married guys did not want to go and I volunteered. No information was given, only that the deployment was classified and might involve combat. I arrived at Korat in a C-130, the second aircraft to arrive. The F-105s were still at Clark. When we got off the Hercules, the facilities consisted of five wooden shacks with tin roofs and lots of empty ramp space. For obvious reasons we called it "Camp Nasty." There was nothing we needed and everything had to come in on C-124s and C-130s.

As to the August 14 mission, Case remembered:

When we got to the target area, our lead, Jack Stressing, contacted the FAC and asked what the target was. He answered it was an Anti-Aircraft [AA] site and that he would fly over it to get them start firing so we could see where it was. When it did, Jack located it and told us to take spacing. He did a 270-degree turn and rolled in. Dave [Graben] followed. Unknown to us was that Rod Beckett had lost his radio, had no idea what was going on, and just followed #2 around the pattern about 3,000 feet in trail. I followed Beckett. Just as I was about to roll in, Dave radioed he was hit and heading home. Lead said he would join up with him and that I was to join #3 after my pass on the target. I rolled in, put the pipper about thirty mills high, squeezed off about 200 rounds of 20mm, fired the 2.75-inch rockets and another 200 rounds. I then pulled off the target, joined up on #3 and Returned to Base, RTB-ed.

Combat loss

Lt Graben's aircraft (62-4371) was hit by 37mm fire, resulting in a 20in x 24in hole in the horizontal stabilizer control surface and 15 shrapnel holes in the vertical stabilizer. He also experienced an in-flight fire, which burned through just forward of the speed brake. The engine was okay, except for some afterburner damage. A note of interest is that the USAF considered the aircraft as a combat loss, while PACAF mentioned that a F-105 had received battle damage. In reality, "371" received a new aft section and was back in action a few days later. It remained in the inventory until it was lost over North Vietnam on September 20, 1966.

A second SAR-supporting mission was flown on August 18, when four pilots, Capts Ted Shattuck (Lead) and Rick Layman (#2), 1Lt Larry Van Pelt (#3), and Lt Col Donovan McCance (#4), were involved in the successful rescue of another Air America T-28 pilot. This time, an Air America C-123 was the FAC. They also used rockets and guns against 37mm gun emplacements. All involved earned the Air Medal. Flying time was three hours and ten minutes. The use of gun cameras during the August 14 and 18 missions proved to be a complete failure, as much of the film never went through the cameras. This meant the loss of the first combat pictures using F-105s against ground targets.

On September 8, 1964, the 36th flew four night sorties. However, the lights for the first 4,000ft of the runway went out, making it necessary to use 35 parked vehicles with headlights turned on to illuminate the unlit portion of the 10,000ft runway. Ten days later, the US Navy reported unidentified boats in the vicinity of its destroyers, resulting at Korat in a five-minute alert posture. Apparently, it was a false alarm, and as the Squadron's history stated, "the 'incident' was soon forgotten and passed into history with the first two incidents." On the 20th, the alert status was reduced to a three-hour posture.

On October 5, the Squadron returned to Yokota WPO (With Personnel Only), with the 35th TFS assuming the commitment. Interestingly, the motto of the 35th is "First to Fight," but the 36th TFS was the first PACAF F-105 unit to see combat in SEA. One of the lessons learned was that training programs were inadequate to suit the type of battle conditions in SEA, preventing the 36th from fully exploiting its potentials. As a result, it was recommended to make training missions more realistic.

And Marty Case? His Bobbie had been waiting for him, and shortly after his return they got married. And as a matter of fact, they still are!

357th TFS

The F-105D squadron in the CASF, deploying to Yokota, was the 357th TFS (355th TFW, McConnell AFB, Kansas). Originally, the Squadron was scheduled for a 90-day TDY to Incirlik AB (Turkey), as *Fox Able 169*, to take over the NATO SIOP commitment from its sister squadron, the 354th TFS, with an August 10 departure date.

The 357th deployed as *One Buck Two*, with older series aircraft. Launch of the first six of 22 aircraft (including four spares), led by Squadron commander Lt Col George Cleary, was initiated at 07:00L on August 6. All legs were flown non-stop, with four air refuelings on the first leg (McConnell–Hickam), three on the second (August 7 to Andersen), and only one to Yokota (August 9). Total flying time was 18 hours and 20 minutes. Not everything went as advertised, however, as, on the first leg, Maj Kalspach aborted into Amarillo AFB (Texas) with serious aircraft problems. He was accompanied by Capt Saffel. The other 20 aircraft proceeded, but, at the first air refueling point, Capt Langwell could not take on fuel and aborted into George AFB (California) in the company of Maj Herman. Langwell's aircraft was repaired, and both pilots launched at 14:00L for Hickam in the company of a single Stratotanker. They landed at Hickam at 16:00L on the 6th. At 06:30L the next morning, the first wave of seven aircraft departed Hickam. The number should have been eight, but the unlucky Langwell aborted on start-up with a leaking fuel line. The other two waves launched with 30-minute intervals for a total of 19 aircraft. On this leg, all pilots logged weather time due to numerous thunderstorms. McCleary's wave landed at Andersen on August 8 at 10:00L. The pilots encountered no aborts or major problems. This was different for the third leg. Only 17 Thunderchiefs launched at 06:30L on the 9th, as Capt Gordon aborted with an air turbine motor (ATM) failure, and Capt May's F-105D experienced a hydraulic leak. The first wave landed at Yokota at 08:30L. Capt May arrived as number 18 on the 10th after the hydraulic leak was fixed. The four stragglers eventually also arrived, by commercial air. While at Yokota, the Squadron was attached to 41AD.

Dave Ferguson was one of the pilots. He recalled:

> We came in with our complete mobility package, absolutely everything needed to operate. We had eighteen F-105Ds with probe & drogue only, twenty-four pilots, maintenance, etc. The reception was fine, a lot of us knew each other. We picked up all the commitments of the squadron that deployed. We were all combat-ready, including all nuke deliveries. After landing at Yokota we immediately began to study our targets, which we were to pick up at Osan. The alert there was one week. We used our own aircraft and generally deployed as a flight. We had three flights, so generally you would spend one week at Osan and two weeks at Yokota. We had six aircraft on nuke alert. The configuration varied with the target. With weapons on the inboards, you usually had a bomb bay tank (390 gallon) and a 650-gallon tank on the center line. Lots of gas! When the nuke was carried internally, you had the 650, which was dropped when dry, and two 450-gallon tanks. During that week, we would normally fly one day. I remember there was a constant swapping of aircraft between Osan and Yokota. In the two weeks at Yokota, we flew normal training sorties, which were pretty much the same as in the US. Most of them were flown with small bombs (Mk-106 "beer cans" and BDU-33 "slicks"). We occasionally flew with concrete shapes. If the shape was carried inboard, we had a 450-tank on the other inboard, so that the aircraft was still near symmetrical and easy to fly.

Aircrews also TDY-ed to Kadena, as they had an AGM-12B Bullpup simulator. Ferguson continued:

> We also used their range to fire the AGM-12B and AIM-9B. Actually we flew to Kadena quite a few times for navigation purposes and to keep up flying time for the 355th, which also had a squadron pulling nuke alert at Incirlik. Usually, when we had any fuel left, we would try to do at least one ACM [Air Combat Maneuvering] engagement, either 1V1 or 2V2. Between 105s only. No dissimilar ACM was allowed.

According to Ferguson, there was a little rivalry through an unofficial competition to see which squadron flew the most. As he recalled:

> We always flew more than the other two squadrons combined and this irritated "the hell out" of the PACAF brass. There were three main reasons, (1) it was shown over and over that a deployed squadron can operate much more effectively without all the cumbersome rules of centralized, "communist," maintenance: Yokota knew centralized maintenance, while we kept our own identity; (2) Yokota had recently moved from Itazuke and had transitioned from the 100 to the 105: our maintenance was much more experienced; and (3) we had the early Ds which had the least problems. We continued to outfly them during our stay.

The 357th returned to McConnell WPO on December 12 and was relieved by its sister squadron, the 469th, as *One Buck 6*. They had sent an advance party over. Between 6–8 pilots were left behind, as the 469th was not fully manned.

Subject to attack by North Vietnamese torpedo boats on August 2 and 4 included the USS *Maddox*. In the photo, the destroyer is operating off Oahu, Hawaii, on March 21, 1964. (USN, PH2 Antoine)

After the August 4 attack on the *Maddox* and the USS *C. Turner Joy*, a retaliatory attack, Operation *Pierce Arrow*, was directed. Aircraft off USS *Ticonderoga* and *Constellation* flew 64 strike sorties. This September 1964 photo shows A-4C 149553 of *Constellation*'s VA-144 taxiing at Da Nang AB, Vietnam. (USN, via Gary Verver)

Augmenting TAC (Tactical Air Command) resources were two F-100D/F squadrons, the 522nd (27th Tactical Fighter Wing [TFW], Cannon AFB, New Mexico) and 614th TFS (401st TFW, England AFB, Louisiana), and six RF-101Cs of the 363rd Tactical Reconnaissance Wing (TRW, Shaw AFB, South Carolina). The aircraft staged through Hickam for crew rest. (USAF)

5AF was directed to send one F-105D squadron to Korat and selected Yokota's 36th TFS (41AD). Capt Ken Furth has just arrived and is deplaning from his F-105D, 62-4362. (USAF)

Flight line security was immediately initiated after arrival at Korat, with A1C William Cisco as living proof. Behind him, F-105D 62-4399/F (squadron color was blue) of the 35th TFS. Note the F-105F in the lineup. (USAF)

Maintenance personnel working in the cockpit of F-105D 62-4329/W (blue). (USAF)

After arriving at Korat, four Thunderchief pilots were put on five-minute cockpit alert to fly SAR (Search and Rescue) support missions over Laos when required. The F-105D is 62-4327. (Vic Vizcarra)

Above: The flight line at Korat with at least 35th and 36th TFS (red) F-105Ds. (Bob Pielin)

Right: The aft section of Dave Graben's F-105D, 62-4371, after landing at Korat on August 14. The USAF categorized "371" as a combat loss. Pilot and aircraft were actually lost on September 20, 1966. (Capt Clarence Fox)

Proof that "371" was still alive and kicking after engaging hostile fire over Laos. It is obvious the aft section received a new tail number. (Capt Clarence Fox)

Left: To help preserve the Single Integrated Operational Plan (SIOP) commitment of the 36th TFS at Osan AB, TAC was directed to send a F-105D squadron to Yokota. The 357th was selected for the *One Buck Two* deployment, and its low-Block F-105Ds are shown prior to take-off from McConnell AFB for the first leg to Hickam. (USAF)

Below: The second leg was to Andersen. F-105D 59-1764 was one of the two (354th TFS) Thunderchiefs claimed on April 4, 1965, by North Vietnamese MiG-17 pilots during a strike against the Thanh Hoa Railroad and Highway Bridge. (Dave Ferguson)

Chapter 2
Temporary Duty at Korat, Da Nang, Takhli and Tan Son Nhut

Korat

When the Thunderchiefs of the 36th TFS arrived at Korat in early August 1964, it was not the first time the base had hosted F-105s. On April 21, 1964, it had done so when 12 F-105s arrived for Tactical Air Exercise *Air Boon Choo*. It was the first time F-105s had participated in a Southeast Asia Treaty Organization (SEATO) exercise. In late March, the 44th TFS had been designated to participate and to carry the entire 18th TFW commitment with aircraft, 20 pilots, 165 maintenance/support personnel and support equipment. The deployment was led by Lt Col Grant Smith, the commander. Redeployment was on May 2–3, and 170 sorties/352 hours were flown.

The USAF organization in Thailand in 1965 was a rather confusing one, with many designations, activations, discontinuations, etc. The arrival in the spring of one F-4C and four F-105 squadrons, plus RF-101, B/RB-66 and other aircraft, demanded a wing-size organization. For this reason, on April 5, PACAF designated TFW Provisional, 6234th and organized it at Korat. Its mission was to exercise operational and administrative control over all USAF units in Thailand, until establishment of permanent wings. Before the Wing's activation, the preexisting units had been administered by the 35th Tactical Control Group at Don Muang. The Provisional Wing was discontinued on July 8 and replaced by the newly established and more permanent 6234th TFW. It was assigned to 13AF and attached to 2AD. However, the Wing had to wait until November to have Thunderchiefs assigned. Major F-105 maintenance was accomplished at Kadena, as designated Main Operating Base (MOB), including phase and periodic inspections.

In late 1964 and in 1965, the 18th TFW at Kadena, and the 355th TFW used Korat during their TDY assignments. In addition, 41AD's three F-105 squadrons deployed to Korat in the August–December 1964 period. The first 18th TFW unit to TDY here was the 44th TFS "Vampires," on December 18. Deployment was named *Blue Mood I*, with the Squadron relieving the 80th TFS with simultaneous attachment to 2AD and becoming part of Det 3, 18th TFW, which was also commanded by the 44th TFS commander, Lt Col Bill Craig. As a consequence, TAC's augmenting F-105 squadron, the 469th TFS, moved from Yokota to Kadena. The 44th deployed as a unit and included 22 pilots and 18 F-105D/Fs. Of these, only one or two were Fs, which were primarily used for instrument checks and admin needs. On occasion, they were used on a combat sortie with a strapped down back seat. Maintenance was performed on an open ramp without protection from the heat and other elements, while Operations was housed in tiny, thatch-covered buildings. The 80th TFS assisted the 44th with familiarization and "settling-in." Through late December, the 80th flew 487 hours in 198 combat (support) sorties. The first increment of six F-105Ds returned to Yokota on the 22nd, the second on the 29th, and the final six aircraft departed Korat on January 5, 1965. While at Korat, the 44th TFS cycled flights of six aircraft to Da Nang off and on for 3–4-day periods to fly C-130B-II and RB-47H escort missions over the Gulf of Tonkin. One of the missions at Korat was Rescue Combat Air Patrol (RESCAP). The Squadron was directed to prepare flights of aircraft for scramble take-offs as and when directed by the Air Support Operations Center (ASOC). The standard posture was four aircraft on one-hour alert, with ASOC having the option to cut launch times to as little as five minutes. While at Korat, seven RESCAP missions were flown with 49 sorties. On February 19–20, 21 F-105s were involved in the RESCAP for Maj Bob Ronca, Commander of the 613th TFS (F-100D). By December 31, 18th TFW aircrews had flown 111 F-105D/F sorties from Da Nang and 103 from Korat, with 265 and 232 flying hours, respectively.

Korat TDY Deployments

Sqn/AD/Wg	Periods
44/18	April 21–May 3, 1964
36/41	August 9–October 5, 1964
35/41	September 24–November 20, 1964
80/41	October 30–December 29, 1964
12/18	February 2–March 15, 1965* and June 15–August 25, 1965
44/18	December 18, 1964–February 25, 1965, April 21–June 23, 1965 and October 19–29, 1965**
67/18	February 8–April 26, 1965 and August 16–October 23, 1965
354/355	March 3–June 12, 1965
357/355	June 12–December 3, 1965*** (was TDY at Yokota, August 9–December 12, 1964)
421/355 469/355	Although these units were on TDY at Kadena (April 7–August 20, 1965 and November 30, 1964–March 13, 1965, respectively), they were able to send flights to Korat for their aircrews to gain combat experience. For instance, in January, the 469th flew 457 hours, of which 299 hours were for training and 158 were for combat/combat support.
Det A 561/23	April 8, 1972–September 6, 1973. Attached to the 388th TFW.

* On February 2, the Squadron deployed to Da Nang as a unit. Between February 8–20, the Squadron was split between Korat and Da Nang. On the 20th, the Da Nang personnel and equipment also moved to Korat.
** Five pilots volunteered to remain at Korat, flying as D flight for TAC's 357th TFS, which was undermanned. They were known as "Murch's Marauders" (Capt Thomas Murch was the flight commander). They returned to Kadena on November 14.
*** Moved from Korat to Kadena on August 28 due to the extensive F-105 modification program and were relieved by the 68th TFS (F-4C) from George, leaving just one TDY F-105 squadron, the 67th, at Korat. By November 5, the 357th gradually returned to Korat, initially with eight aircraft, replacing the 44th TFS. Fourteen days later, the Squadron had 15 F-105Ds. In the meantime, on the 15th, the Squadron F-105Ds at both Kadena and Korat were transferred to the 6234th TFW. The 68th TFS moved to Ubon RTAFB on November 24.

Above: A Thai military band marching at Korat's taxiway during *Air Boon Choo* in April 1964. In the background, F-105Ds of all three maintenance Sections, Yellow (12th TFS), Blue (44th TFS), and Red (67th TFS). (Dag Damewood)

Right: In February 1965, F-105Ds are ready for their next combat mission. F-105D 61-0055 was assigned to the 469th TFS, which was TDY at Kadena AB. (Ed Skowron)

Below: F-105Ds 61-0210 (Blue) and 62-4248 are flown by 44th TFS pilots while deploying from Kadena to Korat in April 1965. (Dag Damewood)

F-105D 61-0197 (18th TFW) in May 1965, on its way to last chance for a final check by maintenance and to arm the ordnance, eight M-117s, by armament personnel. (Ed Skowron)

Armament personnel of the 18th TFW are loading LAU-3/A pods with 2.75in rockets, in August 1965. (Frank Street)

On August 15, 1965, Lt Col Bob Fair, 12th TFS Ops Officer, and Maj Bill Hosmer were en route in F-105F 63-8272 (Yellow) from Korat to Udorn RTAFB for a conference, when they experienced an AC generator and utility system failure. They decided to return to Korat. Touchdown was normal, with the drag chute deploying normally. After 2,000ft of roll, "272" started veering to the right, left the runway, shearing the nose gear and buckling the fuselage. Hosmer jettisoned his canopy and vacated the aircraft without a problem. As the front canopy could not be opened or jettisoned, Fair was forced to chop himself out, using the breaking tool. The aircraft sustained major damage. It was disassembled, crated and shipped to the Continental US (CONUS). Repair was completed on January 10, 1967, for a loss of some 513 days. The photo shows "272" at the Korat flight line. (Frank Street)

F-105D 61-0208 (Red) is returning to Korat after a combat mission. The M-117s were expended, but at least one LAU-3 pod was "saved." (Frank Street)

Above: Later in 1965, when all squadron maintenance was centralized, the nose bands of 18th TFW F-105D/Fs were replaced by a broad yellow tail band, like on F-105D 62-4335, seen here. Armament personnel are loading the Multiple Ejector Rack (MER) with six M-117s. The pod on the left outboard is a movie camera. Note the combat sortie markings in red, meaning all were flown by 67th TFS pilots. (USAF)

Left: Christmas 1964 at Korat, with tree decoration undertaken by 80th TFS personnel. (Bob Pielin)

80th TFS F-105D 62-4360, still with the yellow intake and flap, but without the colored letter on the tail, returns to Yokota from Korat in December 1964. (Bob Pielin)

Above: Although the prime responsibility of the 469th TFS, after its move from Yokota to Kadena, was to assume the SIOP responsibility of the deployed squadron, flights deployed to Korat to gain combat experience. Hence, a 469th TFS aircraft, 60-0417, is in the lineup of 18th TFW Thunderchiefs, in February 1965. (Frank Street)

Right: After the briefing and gearing up, pilots walk across the flight line to their assigned Thunderchiefs. F-105D 59-1761 is a 354th aircraft, as is 62-4302, but yet without squadron markings. (Ed Skowron)

Four 354th TFS F-105Ds in May 1965, just prior to take-off for a combat mission. (Ed Skowron)

Left: In May 1965, the 354th TFS pilot of F-105D 59-1761 is using probe-and-drogue to be refueled from KC-135A 58-0063. (Coll/TvG)

Below: F-105D 62-4325 (357th TFS) at last chance at Korat. (Bill Ramage)

After the introduction of centralized maintenance, the 355th TFW painted the colors of its four squadrons on the tail: blue (354th), yellow (357th), red (421st) and green (469th). F-105D 58-1163 (357th TFS) is configured with MLU-10/B mines. (Bill Ramage)

After the North Vietnamese invaded the South in force in late March 1972, the US hastily reinforced its forces in SEA. One of the units called upon was McConnell's 561st TFS, which was directed to send 12 WW-F/G aircraft with personnel and equipment to Korat. Four of these aircraft are shown at Hickam. The three on the left are ALQ-105-configured G models. (USAF)

Two WW-Fs in May 1972. 62-4428 carries the name *Red Ball* on the intake. Red ball was a term used for an aircraft that developed a mechanical problem, resulting in an abort. "428" red-balled so often, it was nicknamed *Red Ball*. (Richard Kierbow)

On or about July 25, 1972, the Motel Delta tail code of Det A's Thunderchiefs was changed to, officially, Whiskey Whiskey, and, unofficially, Wild Weasel. F-105G 62-4434 is in the company of F-4E 69-7551 (469th TFS). (USAF)

Two F-105Gs of Det A, 561st TFS, and one of the 17th WWS are at the head of Korat's runway just prior to take-off. (Ed Skowron)

Da Nang (see also Chapter 3)

As directed by PACAF OPlan 131-65, Alpha Flight of the 44th TFS deployed on December 4, 1964, with eight pilots and six F-105Ds to test Da Nang's ability to support F-105s. When the Squadron deployed to Korat on December 18, six of its F-105Ds, as well as aircrews, support personnel and equipment, deployed to Da Nang in support of "131-65." They became part of Det 2, 18th TFW. Despite almost non-existent support facilities, the aircraft were turned around immediately to fly *Box Top* RB-47H (the first on December 19) and *Queen Bee* C-130B-II (December 21) escort missions over the Gulf of Tonkin. On the 22nd, the contingent deployed to Korat and joined Det 3. However, just before Christmas, the 44th returned to Da Nang with six F-105Ds, aircrews and maintenance personnel to fly *Barrel Roll #4* on the 25th. By year's end, the Squadron still had six F-105Ds at Da Nang. On January 12–13, the 44th TFS "Vampires" was at Da Nang with 18 aircraft to participate in *Barrel Roll #9*. For the same reason, the 67th TFS sent six F-105Ds and aircrews. Missions flown by the 67th during January 12–18 included *Yankee Team* escort. One such mission, YT 1342 (EW-2) for the RF-101C and YT 6649 ATA for the F-105s, was flown on the 17th. Reconnaissance targets included two military areas, the Ban Seng Phan Highway Bridge and the Ban Lang Khang AAA batteries. Aircraft involved included one RF-101C of Det 1, 33rd Tactical Group (TG) (based at Tan Son Nhut), Fern 54, and two primary F-105Ds for escort, Steer 05 and 06, and two spares, Steer 07 and 08, which were not to penetrate the Laotian border unless replacing primary aircraft. The Thuds were configured with two 450-gallon fuel tanks, two CBU-2s, and a maximum of 20mm HEI ammunition. Weather reconnaissance was flown in advance of the mission and eight F-105s were on ground alert (at Korat) to provide RESCAP when needed. The Voodoo received 37mm and automatic weapons fire, which was not retaliated by the Thunderchief pilots. They expended the CBUs and 425 20mm rounds at targets in the Ban Lang Khang area and at a possible 37mm site. None of the aircraft received battle damage. Seventy-nine hours were flown on combat/combat support sorties while on TDY.

Da Nang TDY Deployments

Sqn/AD/Wg	Periods
80/41	December 13–14, 1964
12/18	February 2–20, 1965 (see Korat)
44/18	December 4–, 1964, December 18–22, 1964, December 24–, 1964, January 12–13, 1965
67/18	January 12–18, 1965
36/41	March 9, 1965

In mid-June 1964, the first F-100Ds arrived at Da Nang. At least five Super Sabres of the 615th TFS/401st TFW are on alert. To the right, C-123B Providers. (USAF)

The first *Barrel Roll* mission in Laos originated from Da Nang and was flown by four F-105Ds of the 80th TFS, which, plus two spares, had deployed from Korat. The pilot of F-105D 62-4370 is taxiing at Takhli during a later deployment. (Marty Case)

Above: On February 1, 1965, the 12th TFS deployed to Da Nang as a unit. The photo shows a busy flight line with at least nine F-105Ds showing the colors of all three maintenance sections. (Arnie Bookheim)

Right: F-105D 62-4279. F-100D 55-2894 was assigned to the 416th TFS. (Arnie Bookheim)

Below: F-105D 62-4271 (Yellow) is about to be taxied out for a combat mission. (USAF)

Takhli

Until the establishment of the 6235th CSG on May 8, 1965, provision of administrative, logistical and housekeeping support for all USAF units assigned, attached, or deployed to Takhli was undertaken by Air Base Squadrons (ABS). On November 8, however, the Group was inactivated and replaced by the 355th TFW.

In 1965, the 41AD/6441st TFW at Yokota (on April 1, the 35th, 36th, and 80th TFS were reassigned from 41AD to the 6441st TFW, which was designated and organized on that date), the 4th TFW (Seymour Johnson AFB, North Carolina), and the 23rd TFW (McConnell) used Takhli during their TDYs. Major F-105 maintenance was accomplished at Yokota, as designated MOB.

On February 26, 1965, 41AD was directed to deploy an F-105 squadron under OPORD 145-65. 5AF Frag Order #1 to "145-65" designated the 36th TFS, which received the relevant warning order, also on the 26th. After several days of preparation, the Squadron deployed on March 4 with 16 aircraft, of which two were spares. Four of them diverted into Kadena, after three were unable to take on fuel at the first refueling. After they were turned around, they departed for Clark, where they remained overnight. In the meantime, the other 12 arrived at Takhli after a 6½-hour flight, requiring two in-flight refuelings. The Clark stragglers arrived on March 5, supported by an Enroute Support Team (EST) that had arrived on the 3rd. Four additional aircraft joined the Squadron at Takhli, two on the 5th and two on the 6th. These F-105Ds diverted from a *Joss Stick* exercise at RAAF Butterworth, Malaysia. On March 20, two F-105Ds were returned to Yokota. Deployment was made with 198 maintenance/support personnel. The already present F-100 units (as of early June 1962, [detachments of] TAC F-100D/F squadrons deployed to Takhli continuously under *Saw Buck* and *One Buck* respectively, generally for 60-day periods) provided the 36th with an up-to-date operational briefing, including all codenames being used to delineate the various types of missions. For the first 11 days, Squadron Ops was housed in the same building as Base Ops, which was physically located one mile from the F-105 parking area. Maintenance was housed in a small hut near the flight line. The first familiarization sorties were flown on the 7th. Two days later, the 36th deployed six aircraft to Da Nang to fly a *Box Top* mission. However, due to bad weather, the aircraft were instead placed on five-minute alert, then returned to Takhli later that day. Additional deployments were made to Da Nang to fly *Queen Bee* missions. These missions involved four aircraft and were approximately five-and-a-half hours in duration with five air refuelings. Each aircraft was configured with two AIM-9Bs on each outboard. The first combat sorties were flown on March 14.

In accordance with a February 1 PACAF message, the resident F-100D/F squadron, the 428th TFS on *One Buck 5*, departed on March 7 for Clark after arrival of the Thunderchiefs, taking along eight F-100D/Fs. Arrival at Cannon was on the 13th. A detachment of the 416th TFS arrived at Takhli from Clark on March 18 to augment the 36th TFS until arrival of the second F-105D unit, McConnell's 563rd TFS, and then moved to Da Nang after the 416th deployed there from Clark on April 22. The "US Aircraft Deployable Forces Inventory" as of April 30, 1965 (part of "Management Summary Southeast Asia" of May 13) was the last one, showing 18 F-100D/Fs at Takhli. It also showed 39 F-105s.

According to the Air Force Historical Research Agency at Maxwell (Alabama), there was also a provisional wing at Takhli, TFW Provisional 6235th, from April 8–July 8, 1965, but it was not manned and never made operational. In general, units were attached to 2AD. When the 355th arrived at Takhli, the attachment of TDY squadrons (35th, 334th, 335th and 562nd TFS) transferred to the 355th.

Temporary Duty at Korat, Da Nang, Takhli and Tan Son Nhut

Takhli TDY Deployments

Sqn/AD/Wg	Periods
36/41AD*	March 4–May 4, 1965, and August 26–October 28, 1965
35/6441	May 4–June 26, 1965, and October 19–November 9, 1965
80/6441	June 27–August 26, 1965
334/4	August 28–February 5, 1966
335/4	November 9–December 6, 1965 (was TDY at Yokota, July 3–c. November 6, 1965)
562/23	August 6–c. December 4, 1965
563/23	April 8–August 15, 1965
561/23	Although this unit was TDY at Yokota (March 6–July 8, 1965), it was able to send flights to Takhli

* As of April 1, 6441st TFW.

Right: Before the 335th TFS deployed to Takhli on November 9, it was deployed at Yokota to perform the SIOP mission at Osan. However, when at Yokota, training missions were flown. Here, two F-105Ds have just touched down after such a mission. (Yukio Enomoto)

Below: Units of the 4th TFW entered the air war relatively late: it was not until August 28, 1965, that the 334th TFS deployed to Takhli. The November flight line shows aircraft of the 334th, like camouflaged 62-4240, and of the 335th in the second row. (USAF)

Left: A flight of four 334th TFS F-105Ds, in company of KC-135A 57-1422, while in the prestrike refueling phase of the mission. (USAF)

Below: A November photo of the flight line with 334th aircraft to the right and of the 335th TFS to the left. The latter lineup also includes one 6441st TFW F-105D, 62-4318. Note that aircraft of both 4th TFW units show TAC and PACAF emblems. (USAF)

Camouflaged F-105D 62-4282 is adorned with the polka dot markings of the 334th TFS. (USAF)

Takhli in April. The F-105Ds in the foreground are of the 563rd TFS, and those in the background are of the 36th TFS. The B-66s are RB-66Cs, which deployed to Takhli as *Two Buck 2*, and were mainly used in an active electronic countermeasures (ECM) role. (Marty Case)

F-105Ds on last chance in April. The aircraft were assigned to the 563rd TFS, which arrived at the base on April 8. They were still in standard TAC markings. (Marty Case)

This August lineup includes F-105Ds of three different 23rd TFW units: 62-4386 shows 563rd TFS markings, the tail of 62-4376 shows the 562nd, while the fifth tail interestingly still shows 561st TFS markings. (USAF)

563rd TFS F-105Ds on last chance in various stages of applied squadron markings, including the Ace of Spades. (Marty Case)

When the 562nd TFS arrived at Takhli to assume the 563rd's responsibility, it also obtained its aircraft, which were then adorned with the blue-white-blue tail markings, like on 61-0090. The pod on the left outboard is a movie camera. (USAF)

On November 16, the pilot of this 562nd TFS F-105D, 62-4266, made it to Da Nang despite the combat damage. He touched down "clean," meaning he not only jettisoned his ordnance, but also fuel tanks and pylons. The repair, by a Rapid Area Maintenance (RAM) Team, was completed on December 23 and took 925 man-hours. (Coll/TvG)

When it appeared the Pathet Lao would reach the Mekong River and perhaps cross it into Thailand from Laos, in May 1962, the 510th TFS/405th FW, deployed its F-100D/Fs from Clark to Takhli. The pilot of F-100D 56-3270 is being directed by his crew chief to come to a full stop. (USAF)

An 80th TFS pilot (yellow helmet) on a May 1965 combat mission in 36th TFS 62-4391. Note the bombing mission markings in the various colors under "force." (Bob Pielin)

This is certainly not the best quality photo, but it is included due its uniqueness. The aircraft involved is F-105D 62-4346 (35th TFS), taxiing at Takhli in May. The markings were very short-lived. (Marty Case)

Above: Before the use of a flight line at Takhli, aircraft were parked on rosettes, making maneuvering quite difficult. The photo shows two F-105Ds of the 6441st TFW and one of the 563rd TFS. 61-0156 is being painted in Ace of Spades markings. (Davidson)

Right: While the 334th F-105Ds, like those in the background, still have their squadron markings in October 1965, the aircraft of the 6441st TFW lost theirs and sport Wing markings, blue, yellow and red band on the tail. (USAF)

Tan Son Nhut

It is possible to configure the Thunderchief with 26 Mk-82 500lb or 16 M-117 750lb bombs. There was a big disadvantage though, as the combat range was limited greatly without external fuel tanks. Of course, air refueling was a possibility, but the risk of losing KC-135s while refueling F-105s on combat missions over North Vietnam was not a preferable option. Did this mean no combat missions were flown in SEA by F-105s configured with a maximum bomb load? There were, and the 563rd TFS had the "honor."

The Squadron, as the first 23rd TFW unit, deployed to Takhli on April 8, 1965, as *Two Buck Five*, with 18 F-105s, 35 officers and 315 enlisted personnel. In the afternoon of June 25, it was directed to deploy six F-105Ds to Tan Son Nhut near Saigon for a "special mission." The aircraft, four primary and two spares, of D, Delta Flight arrived later that day in darkness, followed by two more pilots and maintenance/support personnel and equipment, including extra Multiple Ejector Racks (MERs), in a C-130. After arrival, the aircraft were parked on a taxiway as there were no hardstands. The base also hosted a squadron of F-100D/Fs from Cannon, the 481st TFS. One of the deploying 563rd pilots was Capt Al Logan, who was the Squadron weapons officer. Logan recalled:

> The next morning we flew a flight of four up to northern South Vietnam, configured with eight M-117s each, where we worked with a FAC. The other two pilots flew a similar afternoon mission. In the meantime, the four primary aircraft were loaded for the max load go next day, with MERs replacing the 450 tanks on the inboards. After the intelligence/target briefing, I briefed the flight about tactics, techniques and procedures, as well as sight settings, etc. I specifically briefed to hold the pickle button down long enough for all bombs to come off. There was no air refueling and no strafing runs were to be made after dropping the bombs. About mid-morning the four aircraft were launched. Arriving in the target area in the Iron Triangle, Mekong Delta, south of Saigon, the O-1 FAC was contacted, who marked the Vietcong targets with smoke. After the strike he remarked he was impressed with the results. The mission lasted 35–40 minutes.

Al Logan believed another mission with a normal bomb load was flown later that day, this time with all six aircraft involved. Everybody then returned to Takhli. On August 14, 1965, operational control was assumed by the 562nd TFS, which had deployed WPO. The next day, the 563rd initiated redeployment to McConnell, completing it on the 17th. The 563rd paid a heavy price, as it lost ten F-105Ds, with only four pilots being recovered.

F-100 armament personnel of the 481st TFS load the final M-117 onto the outboard of 62-4398. (Capt Dick Moore)

A revetment with two F-105Ds, loaded with a maximum load of 16 M-117s. (Capt Dick Moore)

The pilot of F-105D "398" taxiing at Tan Son Nhut AB on June 26 to the runway head to launch a history-making F-105 max-load combat mission. (Capt Dick Moore)

Three of the four participating F-105Ds after take-off on their way to the Iron Triangle in the Mekong Delta, south of Saigon. (Capt Dick Moore)

F-105Ds 62-4398 and 61-0127 on their return trip to Takhli. (Capt Dick Moore)

Chapter 3
First Blood

In November 1964, Laotian Prime Minister Souvanna Phouma was agreeable to air attacks on North Vietnamese infiltration routes and targets in his country. As a result, broad guidance for the expanded program of air operations in SEA was provided in draft National Security Action Memorandum 319 of November 29 and further modified in a conference on December 1 with the US Ambassador to South Vietnam, Maxwell Taylor, the National Security Council, and other top officials. The next day, President Johnson authorized very limited and highly controlled measures to exert more pressure on North Vietnam. It was to be a two-phase program: (I) armed reconnaissance (AR) in Laos for 30 days to begin on December 14. Also, the strike program by T-28s of the Royal Laotian Air Force (RLAF) would be augmented; and (II) a gradually increasing program of military pressure was to begin, as reprisal for "significant" incidents. Planning and force preparation for Phase II were authorized. Phase I was nicknamed Operation *Barrel Roll* (*BR*), and included US strikes on infiltration routes and facilities in the Laotian corridor and AR. Its prime purpose was a psychological one, to signal to Hanoi the danger of deeper US involvement. Initially, however, the program had several restrictions. For example, the number of aircraft to be employed on any one strike mission was limited to four, a period of 72 hours was required between missions, and use of Thai bases for strike aircraft (not authorized by the Thai government) and overflight of North Vietnam were prohibited. On December 11, 1964, the JCS sent instructions to Admiral U. S. Grant Sharp, CINCPAC, to fly two armed reconnaissance/briefed strike missions in the Lao corridor. CINCPAC, in turn, notified its appropriate component commands a few hours later through an execute message, specifying USAF forces under Military Assistance Command, Vietnam (MACV) which would fly the first mission on December 14. 2AD transmitted the frag order on December 13.

Barrel Roll #1 and *#4*
BR #1 was flown on December 14 against targets of opportunity on Routes 8 and 12 by four F-105Ds of Yokota's 80th TFS. To fly the mission, six Thunderchiefs, including two spares, and support personnel deployed to Da Nang the previous day. Capt Jack Redmond later reflected:

> In the morning we flew a two-hour training mission from Takhli and later that day we flew to Da Nang, a direct route right across Laos. Our aircraft were configured with two 450-gallon tanks and no tankers were required. It took us 1hr and 15mins flying time, of which one hour VFR [Visual Flight Rules]. After arrival, the aircraft were turned around and armed. Four aircraft flew the mission with Capts Arthur Means as Lead, John Atkinson as #2, Gordon Walcott as #3 and Dave Groark as #4. Capt Murphy Jones and I lost the coin toss, with me ending up in the tower and Murphy in Mobile Control. The weather was not good, 3-4,000 feet overcast. The four Da Nang-based F-100Ds that flew cover were above the clouds and never saw the F-105s.

In addition, four F-100Ds were on RESCAP, two Det 1, 33rd TG, RF-101Cs accomplished the post-strike Battle Damage Assessment (BDA), while two F-105Ds from Korat acted as Voodoo escorts, while an HU-16B Albatross acted as rescue control aircraft. When the aircraft reached the Nape Bridge, a vehicle was observed on it, and one of the F-105 pilots dropped his six M-117s, which impacted in the water. Arriving at the secondary target, a military strongpoint with several deserted gun positions, 18 M-117lb bombs, one AGM-12B Bullpup (which missed because of low clouds) and 114 2.75in rockets were expended. Results were slim, as the heavy load led to miscalculations of time and distance. Short of fuel, the F-105s made a hurried attack on the bridge, missing it. No enemy reaction was encountered. The aircraft returned to Da Nang, were turned around, and all six returned to Korat in the afternoon in a one hour and 45 minute flight, of which 30 minutes was VFR.

BR #2 was flown on December 17 by 14 US Navy aircraft, including A-1Hs, F-4B Phantoms, and A-4Cs. The primary mission was AR of Routes 121/12, plus the Ban Boun Boa Bridge. The bridge was not damaged, although eight buildings were destroyed.

Just before Christmas 1964, the 44th TFS deployed aircraft to Da Nang. The "Vampires" got its first real action on the 25th, when four F-105Ds, led by Lt Col Craig, flew *BR #4*. The primary target was AR of Route 23, with Tchepone Barracks as the secondary target. Craig's flight was supported by eight MIG/RESCAP F-100Ds, three BDA RF-101Cs, two F-100Ds and two F-105Ds for weather reconnaissance and two KC-135A tankers. Two Det 1, 33rd TG, Voodoos acted as pathfinders for Craig's flight. The secondary target was struck by nine M-117s and 36 CBU-2A pods, resulting in one building being 80 percent destroyed and seven severely damaged. Only light small arms fire was encountered.

Barrel Roll #9

Early January 1965, a CINCPAC representative and US ambassadors Maxwell Taylor, William Sullivan (Laos) and Graham Martin (Thailand) attended a SEA Coordination Committee meeting. To increase the effectiveness of air operations in Laos, it was suggested that an air strike against the Ban Ken Bridge be necessary. Located on the eastern end of Route 7, near Ban Ken Village, the bridge, over the Nam Mat River, served as a main approach from North Vietnam to the Plain of Jars area and was virtually impossible to bypass. This section of Route 7 provided a good highway for the movement of personnel and supplies in support of Pathet Lao military operations in Laos. Back in August 1964, MACV had already considered the bridge a vital interdiction target and suggested to CINCPAC that its destruction could effectively disrupt the flow of traffic. On August 19, CINCPAC agreed and suggested it be hit by Navy aircraft. However, MACV recommended using F-100s and F-105s armed with CBU-2A pods, napalm for flak suppression, and AGM-12Bs. If it was completely destroyed before the Bullpups were expended, the unused weapons were to be expended at surviving AA positions. In an October 6 message to CINCPAC, MACV again recommended the bridge as a desirable target. "Washington" finally authorized the strike and *BR #9*, directed at the bridge, on January 13, became the first important test of the new interdiction program. In July 1964, RLAF T-28s had flown three unsuccessful missions against it with the loss of one aircraft, while a second was damaged.

The limitation of four strike aircraft on *Barrell Roll* missions was lifted for the first time on January 13, when 18 F-105Ds, including two spares, were used. According to Col Craig, the first frag for the mission was probably received on January 9. The next day, he, accompanied by Maj Larry Guarino, his Ops Officer, flew to Da Nang to begin planning the mission. To fly it, the 18 F-105Ds of the 44th TFS deployed with support personnel to Da Nang on January 12. To have enough aircraft available, six additional F-105Ds were flown in from Kadena by pilots of the 67th TFS, led by its commander, Lt Col Robbie Risner. Their TDY Order, T-24, of January 8, stated its purpose was, "In support of PACAF OPORD 131-65." Capt Wes Schierman was one of the pilots, and he stated the following:

The flight was direct going, hitting a tanker in the Philippines area on what was thought to be a "ferry" mission. First impression of Da Nang was not good. We arrived in a thunderstorm and the GCA was misaligned with the runway. We all missed our approaches and had to circle to make a visual landing.

It took me two passes to get on the ground. But we made it down okay, though low on fuel and other options. Other than that, I don't remember any issues. In those days, if a base had a place to park the aircraft, somewhere to hang your chute and helmet, a bed, somewhere to eat and a bar, we were pretty happy. If it just had a bar, we were just happy. After landing we found out a strike mission was planned.

The logbook of Col Cletus Pottebaum, the Wing's Deputy Commander for Materiel, showed that 24 F-105Ds were available at Da Nang at 06:00L on January 13: nine of Blue Section, eight of Yellow and four of Red Section, plus three 469th TFS (TAC) aircraft. With two squadron commanders present, who both felt they could do the job best, the important question of who was to be mission commander had to be answered first. Col Craig "drew the long straw." He noted in this respect, "As to the so-called difficult command situation that occurred during our deployment, I have always believed there were very, very few people that knew about this incident. I never discussed what occurred with anyone, except certain Squadron personnel who were cognizant of what was going on."

At 06:30H(awaii), Craig briefed the participants. Scheduled take-off was at 12:45H. However, after engine start, the mission was rescheduled for two hours later, due to unfavorable weather in the target area as reported by weather reconnaissance aircraft. First take-offs were at 14:45H. The first section of eight F-105Ds (all 44th TFS pilots) was led by Col Craig, the second section by Col Risner, with six 67th TFS pilots, augmented by Maj Larry Guarino and Capt Al Vollmer of the 44th. Six of Risner's aircraft were configured with six M-117s on the centerline MER and an AGM-12B on the outboards, while all other Thunderchiefs had eight M-117s each. In addition, all F-105s had 450 tanks on the inboards. They were supported by 12 F-100Ds, from Da Nang, of which eight were for flak suppression – there were reportedly 45 AW/AAA emplacements protecting the bridge, and previous *Barrel Roll* missions were directed to avoid the area – and four for MIG/RESCAP. The eight 613th TFS F-100Ds were led by Maj Bob Ronca, the commander. His aircraft were configured with AGM-12Bs, CBU-2A pods and their guns. (A little over a month after the January 13 strike, Ronca was shot down and killed. His target? The AW/AAA sites at the Ban Ken Bridge.) The four MIG/RESCAP F-100Ds were assigned to the 428th TFS, configured with AIM-9B Sidewinders and led by Squadron commander Lt Col Ben Clayton and deployed to Da Nang from Takhli. In addition, two RF-101Cs accomplished BDA and acted as pathfinders (Det 1, 33rd TG), while eight KC-135As refueled the F-105Ds. After take-off, the Voodoo pilots led the aircraft to the Nakhon Phanom area in Thailand, just across the border from Laos, where they were refueled. They then proceeded to the target area. Mountainous terrain made approaches difficult, especially with the addition of AW/AAA fire. Craig's section, flying in loose trail, struck the bridge dropping its eastern span and damaging other sections. Risner's section struck the other spans, and severely damaged its abutments and footings. Each F-105 pilot made a successful single bomb run, while six of the pilots had to make two additional runs each to expend their AGM-12Bs. 613th TFS pilots were also required to make multiple passes against the gun emplacements. Due to dust and smoke created by the bombs, Bullpups could not be guided accurately, and only a few missiles impacted on target. After striking their targets, six of the eight 44th TFS aircraft recovered at Korat, while two, plus the two 67th flights did so at Da Nang. The six 67th TFS F-105Ds flew for 14 hours and 20 minutes, and the ten 44th TFS aircraft, 27 hours. Risner and his pilots remained at Da Nang until the 18th, when they returned to Kadena.

Although the bridge was completely destroyed, a bypass was constructed within three days at a nearby ford, permitting the flow of traffic to resume. However, the success had a price. Heavy ground fire had downed one F-105D and one F-100D. In addition, three flak suppression F-100Ds and one RF-101C were damaged, all while at low altitude. One of the 428th F-100Ds was severely damaged, forcing the pilot to make an emergency landing at Udorn, where it was repaired. The pilot of the F-105D (62-4296 of Yellow Section), Capt Al Vollmer, was #2 in Risner's second flight. He was on his second low-level pass to launch his second Bullpup. After being hit, Vollmer was able to fly another 25 nautical miles before having to eject. He was recovered by an Air America UH-34D and flown to Udorn. He had suffered a neck injury.[1] It was the first combat loss of an F-105 in SEA. Positions #3 and #4 in Craig's flight were Capts Ed Skowron and Will Snell. As Ed remembered:

> After Al had ejected successfully, we flew over his position. As we were low on fuel, I marked it on my Doppler and returned to Korat. After landing, our flight leader, Capt Bob Lines, asked if I could find Vollmer again, which I confirmed. Our aircraft were turned around and by the time Bob and I were back at Al's position, he was gone. Unknown to us he had been recovered in the meantime! Flying time was two hours and five minutes.

The lost F-100D, 55-2861, was one of the eight flak suppression aircraft and flown by Capt Charlie Ferguson. He was hit on his fifth pass. He ejected successfully and was also recovered by an Air America UH-34D crew. An Air America C-123B acted as airborne controller in both recoveries. While the pilots were on the ground, RESCAP was provided by the four F-100Ds and four F-105Ds. RF-101C pilot Al Parks had his canopy shot off while making his final photo pass over the target. He made it to Da Nang but flamed out on the runway. The approximate dollar operating cost for the mission was US$4.4m, including the price of one F-105D ($3m) and one F-100D ($1.1m), but excluding repair of the damaged aircraft.

Pilots of two 15th Tactical Reconnaissance Squadron (TRS) RF-101C Voodoos acted as pathfinders for the 16 F-105Ds. The photo shows a Voodoo with pilot and maintenance personnel of the Squadron at Tan Son Nhut. (USAF)

---

1 Vollmer was shot down a second time, on August 17, 1967, while assigned to the 388th TFW at Korat. He was able to get "his feet wet" and was rescued from the Gulf of Tonkin by the crew of an HH-3E. He suffered a badly broken leg and was taken to the hospital ship USS *Repose*.

According to Col Pottebaum, F-105D 61-0204 (Blue) was flown by Lt Col Robbie Risner, 67th TFS commander. Six of his aircraft were configured with two AGM-12Bs on the outboards and six M-117s on the centerline MER. (Coll/TvG)

Capt Bob Lines and crew chiefs of the 44th TFS at the Da Nang flight line, waiting for the launch call. Although it is a Red Section aircraft, it sports a 44th TFS emblem under the cockpit. It is also adorned with "Vietnam ANG," an RF-101 silhouette and a kangaroo stencil. (Ed Skowron)

The two F-100D flak suppression flights of the 613th TFS, led by its commander, Maj Bob Ronca, are taxiing out for *Barrel Roll #9*. (Ed Skowron)

The two flak suppression flights taking off on their way to the target area, the Ban Ken Bridge. (Ed Skowron)

The six 44th TFS pilots who recovered at Korat after the strike are seen in front of their operations "building." L–R: Capts Ed Skowron, Bob Lines, Will Snell, Bob Green, Dean Vikan and Tom Murch. (Ed Skowron)

Capt Al Volmer with neck brace at Korat after his recovery from Laos. The second person is the flight surgeon, Capt Jerome Unatin. (Jerome Unatin)

Chapter 4
Permanent Units at Korat Royal Thai Air Force Base

In November 1961, there were only two small USAF detachments in Thailand, both at Don Muang. In the April–July 1962 period, additional detachments were established at Korat, Takhli and Ubon. As to Korat, in April 1962, one officer and 14 airmen were temporarily assigned as the Joint US Military Advisory Group. By mid-May 1964, three of the six jet-capable airfields in SEA were situated in Thailand – Don Muang, Korat and Takhli. Although the last two bases had a runway and (minimal) support facilities, they had no extensive housing, hangars, shops, etc., needed for the proper operation of a modern jet-equipped tactical wing.

The first unit to be established was Det 1, 6010th TG, on July 10, 1962. On July 8, 1963, it was replaced by Det 1, 35th TG, which, in turn, was replaced on March 25, 1965, by the 6234th ABS to support the TDY fighter units and their operations. The latter was replaced on May 8 by the 6234th CSG. At the start of 1964, the Royal New Zealand Air Force maintained two Bristol Freighters at Korat. In December, construction began on 1,000ft overruns of the main runway, followed by an ammunition storage area and more adequate facilities such as officer quarters, dormitories and wing operations. As stated in Chapter 2, the 6234th TFW was designated and organized on July 8, 1965. Its first fighter squadron, the 469th TFS, PCS-ed WPO from McConnell to Korat over the course of November 9–17 and was assigned on November 8, replacing the 357th TFS, which returned to McConnell. This was followed on the 20th by the PCS and assignment of the 421st TFS, which also deployed from McConnell, WPE (With Personnel and Equipment) as *Two Buck 23B* with 20 F-105Ds. The aircraft were followed, with a 30-minute separation, by four WW I F-100Fs. The 421st was the only McConnell squadron to PCS WPE.

As the first step towards its organization at Korat, the 388th TFW was activated on March 14, 1966, and assigned to PACAF. Ten days later, PACAF published its Special Order G-86, organizing the Wing and assigning it to 13AF with attachment to 7AF, effective from April 8. The Wing absorbed personnel and equipment from the 6234th TFW, which was discontinued through G-76 of March 23. Both fighter squadrons were reassigned to the 388th TFW through G-101 of April 5.

Eventually, Korat also became a MOB, but much later than Takhli: by November 1966, the 388th TFW depended on Kadena for jet engine support only and became independent in that area by April 1967.

PCS F-105 Squadrons at Korat

Unit	Equipment	From	To	Remarks
469TFS	F-105D	Nov 8, 1965	Oct 31, 1972	to F-4Es in Nov 1968; inactivated
421TFS	F-105D	Nov 19, 1965	Apr 25, 1967	reass 15TFW
34TFS**	F-105D	Jun 14, 1966	Dec 23, 1975	to F-4Es, May 1969
13TFS*	F-105D/ F-105F WW III	Jun 23, 1966	Nov 15, 1967	reass 432TRW

Permanent Units at Korat Royal Thai Air Force Base

Unit	Equipment	From	To	Remarks
44TFS***	F-105D/F-105F WW III	Apr 25, 1967	Oct 15, 1969	reass 355TFW
Det 1, 12TFS	F-105F	Sep 18, 1970	Oct 31, 1970	inactivated
6010WWS	F-105F/G	Nov 1, 1970	Dec 1, 1971	inactivated
17WWS	F-105F/G	Dec 1, 1971	Nov 15, 1974	inactivated

*/** Both Squadrons were activated as part of Operation *Spotlight 66*. The 13th TFS was constituted by a May 2, 1966, order from the Secretary of the Air Force (SECAF). The Squadrons, having been activated by a May 2 Department of the Air Force letter, were organized at Korat effective from May 15 through PACAF G-150 of May 10. The 13th TFS was assigned to the 18th TFW and the 34th to 41AD, and both were attached to the 388th TFW for operational control, and administrative and logistical support.

* The 13th TFS was 90 percent manned and equipped from the 44th TFS, with the remainder coming from other 18th TFW units. It would take until June 23 for support personnel to depart for Korat via C-130, while the F-105Ds departed the following day.

** At Yokota, the organization of the 34th TFS was accomplished and manned by reducing the three F-105D/F squadrons from 24 to 18 UE (Unit Equipment). Personnel and supporting equipment were provided from 6441st TFW resources. Eighty percent of the 339 airmen and all 34 officers were volunteers. Yokota's F-105Ds departed for Korat on June 14.

*** Transferred PCS/WOPE (Without Personnel or Equipment) through PACAF Movement Order (MO) #6, April 6, 1967. As "attrition squadron" for the 18th TFW, the 44th had reached a "1/1" status (one officer and one airman) in March 1967. It replaced the 421st TFS.

6234th TFW pilots "fighting" in December 1965 to fly this combat sortie. Prior to their first combat sortie, not yet having experienced what the North Vietnamese would have in store for them? Note the WW I F-100F Super Sabre in the background. (via Jerry Arruda)

Above left: The pilot of F-105D 61-0051 of the 6234th TFW is checking his ordnance, MLU-10/B mines, prior to flying his February 13, 1966, mission. This configuration was quite rare. The aircraft still lacks the APR-25/26 Radar Homing and Warning (RHAW) equipment. (Bob Krone)

Above right: Going home! As can be surmised by the extended receptacles, three 6234th TFW pilots have finished their 100th combat sortie over North Vietnam in April 1966 and are eligible to return to CONUS: Maj Jim Jones and Capts Bud Millner and Bill Secker. The non-camouflaged F-105D, 60-0429, has had the gun access door replaced. (Bob Krone)

At least six F-105Ds made it to Da Nang in May 1966. All aircraft are "clean," meaning their pilots jettisoned all external stores, probably because they were engaged by North Vietnamese MiG-17s. (USAF)

Right: Maintenance personnel make frantic attempts to push a J-75 engine into an F-105D. Especially early on, long hours were made under very warm and humid circumstances. (USAF via *Air Force* Magazine)

Below: The pilot of F-105D 62-4366 *Thunder Blunder* (34th TFS) is starting his J-75 engine in August 1966 via cartridge start. Not a very healthy working condition! (USAF)

Above: WW-F 63-8298 (13th TFS) on November 9, 1966, taxiing in after flying a combat mission. The CBU is still on the left inboard, but the AGM-45 Shrike on the outboard was expended. *Little Angel* is on the front wheel door, while *Sam Dodger* is under the intake. (Don Ayer)

Left: Three 388th TFW F-105Ds are en route to their targets in April 1967. *Mr. Toad* is F-105D 59-1749 of the 469th TFS. It also carried *Marilee E* on the left side. (USAF)

Miss T, F-105D 60-0497 (44th TFS, blue canopy rail and radar reflector) is on last chance on May 5, 1967. The second D on the line is 60-0409 *Bunny Babe* (469th TFS, green). Maj Mo Seaver of the 44th TFS, as Kimona 1, shot down a MiG-17 with "497" on May 13, 1967, as confirmed by the USAF. However, according to the book *Historic Confrontations*, published by the People's Army Publishing House, no losses were acknowledged that day. (John Rehm)

Right: A confusing photo! F-105D 62-4395 *Emily* is assigned to the 421st TFS (Korat), but the maintainer is wearing a 357th TFS (Takhli) cap. Given the background of the photo, it was taken at Takhli. (Lonnie Gillespie)

Below: On June 3, 1967, Capt Larry Wiggins (469th TFS), flying as Hambone 03 in F-105D 61-0069, claimed a North Vietnamese MiG-17 with a combination of an AIM-9B and 376 rounds of 20mm ammunition. It was confirmed by the USAF. The North Vietnamese acknowledged the MiG-17 Wiggins downed was flown by #2 in a flight of four. (USAF)

F-105D 61-0205 *Mr Blackbird* (34th TFS) is taxiing to last chance on June 5, 1967. Under its nose are U-8s of the US Army detachment. (Nick Donelson)

A flight of four F-105Ds during prestrike refueling by KC-135A 57-1514. (Nick Donelson)

The pilot of this 34th TFS F-105D, 60-0424 *Miss Carriage*, expended the bombs from his MER, but supposedly did not engage any North Vietnamese MiGs while on a September 1967 combat sortie. However, Maj Ralph Kuster (13th TFS) flew "424" when he bagged a MiG-17 on June 3, 1967. (Nick Donelson)

That the Thunderchief was a "toughie" was shown, again, on December 12, 1967. While on a *Rolling Thunder 57A* strike against Joint Chiefs of Staff (JCS) Target 9.10, Kep Airfield, Hatchet flight was engaged by North Vietnamese MiG-21s. East-north-east of Kep, #4, Capt Doug Beyer's F-105D 60-0512 (34th TFS), was hit by a probable Atoll air-to-air missile and damaged extensively, causing Doug to jettison his MER with six M-117s. Hatchet Lead, Maj Sam Armstrong, then escorted him to Da Nang, where both landed safely some two hours and 48 minutes after take-off. Repair, by a RAM Team, was completed on February 11, 1968. (via Doug Beyer)

Above: Not everything always goes as advertised. On June 26, 1968, Maj Bryant Heston of the 469th TFS, "borrowing" F-105D 58-1150 (34th TFS), experienced problems on landing when his right main gear axle failed. In addition, Heston lost utility hydraulic pressure, losing control to keep the aircraft on the runway. "150" was destroyed, but Heston escaped uninjured. (USAF)

Left: The pilot of F-105D 60-0518 (34th TFS) flying towards his North Vietnamese target. The Squadron's color was black. (USAF)

Below: An interesting 388th TFW gaggle while on post-strike refueling: two F-105Ds of the 34th TFS and two Combat Martin F-105Fs of the 44th TFS. (USAF, John Evans)

Above: On October 28, 1968, 34th TFS's F-105D 60-0435/JJ completed 100 consecutive abort-free sorties. It was believed to be a USAF record. The aircraft is on last chance and configured with four CBUs on the MER. The nickname *Satisfaction* is still partially visible. (USAF)

Right: *Queen of the Fleet* adorns the fuselage of F-105D 61-0206 (469th TFS) in February 1968. (Jon Alquist)

Below: F-105D 60-5376 (469th TFS, JV) undergoes prestrike refueling. (USAF)

F-105Ds 62-4359 *Cong Buster* (469th TFS) and 59-1760 (34th TFS) on Korat's runway in February 1968, just before the pilots would kick in the afterburners for take-off. (Jon Alquist)

On October 30, 1968, 18 days before the arrival of its new F-4E Phantoms, the 469th TFS flew its 40,000th combat sortie in SEA since arriving at Korat in November 1965, the most of any tactical fighter squadron in Thailand. The flight of four F-105Ds was led by Lt Col Vic Hollandsworth, flying F-105D *Queen Bea*. Three days earlier, he was succeeded as commander by the photographer, who, in turn was succeeded on November 18 by Lt Col Edward Hillding, who led the F-4Es to Korat. (Lt Col James Broussard)

WW-F 63-8285 (44th TFS, *Porky Pig*) on finals to Korat on May 15, 1968. (Coll/TvG)

Armament personnel are loading 44th TFS F-105D 62-4360 with a M-118 3,000lb bomb on the left inboard. (Lee Griffin)

When Korat's runway was closed for major repair, its 40 F-105D/Fs flew from Takhli and its 20 F-4Es from Ubon. On January 29, 1969, the first Thunderchiefs arrived at Takhli after flying a combat mission from Korat, while the opposite occurred on March 1. The Takhli flight line shows F-105Ds of the 34th (JJ) and F-105Fs of the 44th TFS. (Lee Griffin)

The goodbye ceremony for the 44th TFS, which PCS-ed to Takhli on October 15, 1969, and was reassigned to the 355th TFW. (Coll/TvG)

Above: F-105F 63-8311 was one of the six Wild Weasel F-105Fs (WW-Fs) transferring from the 355th to Korat in September 1970. The crew of "311," uncoded, is deplaning after the hop from Takhli. (USAF)

Right: F-105G 63-8292 of the 6010th WWS, configured with an AGM-45 on the right outboard. (Coll/TvG)

F-4E Phantoms of the 388th TFW carried a tail code, starting with Juliet, the 34th JJ and the 469th TFS, JV. When the six WW-Fs arrived at Korat in September 1970, they were assigned to Det 1, 12th TFS, and its parent unit at Kadena carried the tail code Zulu Alpha. Accordingly, Det 1's F-105Fs obtained Zulu Bravo. When the 6010th WWS replaced Det 1, and later the 17th WWS replaced the 6010th, the tail code remained ZB until June 1972, when it became Juliet Bravo (JB). The photo shows a 388th TFW gaggle undergoing prestrike refueling with one F-105F, two F-4Es of the 34th and one of the 469th TFS. (Don Logan)

F-105Gs Wild Weasel (WW-Gs) 62-4442 of 17th WWS, JB, and 63-8351 of Det A, 561st TFS. "442" is configured with an AGM-78 Standard ARM on the right inboard and dual AGM-45 Shrikes on the right outboard. This load came very close to the maximum take-off capacity. According to one of the 17th pilots, Jimmy Boyd, it was a "shitty" load with too much drag, which made it difficult to gain air speed and keep it. (Ed Skowron)

Left: Two F-105Gs of the 17th WWS while on pre- or post-strike refueling in May 1972. (Richard Kierbow)

Below: Two F-105Fs of the 17th WWS are in the de-armament area after completing a combat mission. It looks like the crew of the second F completed their 100th combat sortie over North Vietnam, as can be seen by the extended receptacle. Time to go home! (Lucky Ekman)

At least 11 F-105Gs can be seen on Korat's flight line. (Ed Skowron)

F-105G of the 17th WWS leaving last chance for a training sortie, as shown by the training AGM-45 on the left outboard. (Coll/TvG)

Chapter 5
Permanent Units at Takhli Royal Thai Air Force Base

On June 15, 1965, TAC announced that of its 273 F-105 aircrew members, 175 (64.1 percent) were TDY, of which 28 were in CONUS and 147 in PACAF or USAFE. For the average fighter, this percentage was 48.3. On July 12, TAC had 437 aircraft (including 76 F-105s) of its entire force of 1,373 aircraft on TDY outside CONUS (32 percent).

As it did not look like the situation in SEA would be settled soon, a solution was needed to have units stationed there on a more permanent basis. After authority was received from the USAF to alert units of a proposed PCS deployment, TAC sent a message to 9AF and 12AF on August 3, which served as a warning order for a planned movement of units. The authority was for planning purposes only, and it was emphasized these plans were tentative. MOs would be issued when authorization by higher authority was obtained. Guidelines used in determining this proposed deployment included, among others, that: (1) TAC was to deploy five TFW/HQs, three to South Vietnam (3rd, 12th, 366th) and two to Thailand (8th and 355th); and (2) PCS moves would be one year without dependents. On September 3, TAC published a draft of its Programming Plan 167-65 "Deployment of Units to SEA" as the initial vehicle (Phase I) to provide program guidance for the PCS of five TFW/HQs, two Troop Carrier Wing/HQs, 23 TFSs, eight Troop Carrier and three Tactical Reconnaissance Squadrons (TRSs), and 14 associated maintenance squadrons from TAC to PACAF. Execution was initiated on October 20 and completed in mid-March 1966. A total of 11,807 personnel and 570 aircraft were transferred.

The 355th TFW was notified on October 2 of its move to Takhli, as *Two Buck 23*. One week later, TAC issued MO #17, followed on the 13th by MO-23, ordering the Wing to PCS. On the 20th, all remaining F-105Ds not scheduled to deploy were transferred to the 23rd TFW at McConnell. An ADVON (advanced) party including Col William Holt departed on October 28, arriving on November 1. Col Holt then reassumed the position of Wing commander. One week later, the organizational transfer from TAC to PACAF was complete, and the Wing reassigned from PACAF to 13AF and attached to 2AD, per PACAF G-187 of November 2. Also on November 8, the Wing assumed operational control of the TDY squadrons at Takhli. The November 11 Air Order of Battle (AOB) included 53 F-105Ds (some 23 of which had been transferred by the 6441st TFW), two HH-43Bs and six KC-135As. The Royal Thai Air Force had 16 F-86Fs and four T-6 Harvards.

PCS F-105 Squadrons at Takhli

Unit	Equipment	From	To	Remarks
354TFS	F-105D/F	Nov 27, 1965/WPO*	Dec 10, 1970	to 13AF/WOPE
333TFS	F-105D/F	Dec 6, 1965/WPO**	Oct 15, 1970	to 23TFW/WOPE
357TFS	F-105D/F	Jan 29, 1966/WPO***	Dec 10, 1970	inactivated
44TFS	F-105D/F	Oct 15, 1969/WPE	Dec 10, 1970	to 13AF/WOPE

* Replaced the TDY 562nd TFS, which returned WPO to McConnell.
** Deployed as *Two Buck 29*, replacing the TDY 335th TFS, which returned WPO to Seymour Johnson.
*** Deployed as *Two Buck 23D*, replacing the TDY 334th TFS, which returned WPO to Seymour Johnson.

Above left: This pilot and Thunderchief made it to friendly territory on July 23, 1966. While flying in marginal weather, three F-105Ds, led by WW-F 63-8338 (354th TFS) as Drill 01, ran into a barrage of four surface-to-air missiles (SAMs). The F was hit and went down. The crew was initially listed as MIA (missing in action), but this was later changed to KIA (killed in action). The subject of this photo, F-105D 58-1151 as Drill 02, was exposed to the blast and shrapnel from a proximity burst close to the left side of its fuselage. The pilot made it to Udorn despite sustaining injuries to his left hand and leg, while "151" received 87 holes in the fuselage. The blast also blew off the ventral fin and the top of the fin and rudder. The photo shows the pilot, Capt Buddie Reinbold, who was on his 8th "counter." At Udorn, "151" was prepared for a one-time flight to Air Asia at Tainan, Republic of China, which was completed on September 8 with 1,242 man-hours spent. At Air Asia, another 3,000 man-hours were needed to complete the repair. (USAF)

Above right: A2C William Puthoff of the 354th TFS cleans the canopy of his F-105D, 62-4262, prior to a combat mission in October 1966. (USAF)

An armament technician is checking the wiring of one of the six M-117s on the MER of this F-105D. (USAF, Ken Hackman)

Left: Armament personnel service and load the M-61 Vulcan cannon with 20mm ammunition, while M-117s are still to be loaded. In November 1966, Col Sam Hill was the Wing's Deputy for Operations and attached to the 357th TFS for flying. (Lonnie Gillespie)

Below: This WW-F, 62-4445 (333rd TFS), made it to a friendly base, Udorn, as well. AAA caused the damage on April 23, 1967. Repair was completed on May 6 by a RAM Team. (via RAC/CAM)

The crew chief of F-105D 58-1169 (357th TFS) is helping the pilot get strapped in. (Lonnie Gillespie)

Above: WW-F 63-8277 (357th TFS, yellow) is taxiing on March 14, 1967. The aircraft had arrived on November 26, 1966, and a SAM bagged it on April 26, 1967. (Lonnie Gillespie)

Right: F-105D 59-1731 (357th TFS) is seen while undergoing prestrike refueling. The name on the canopy rail is of Lt Col Arthur Dennis, commander of the 357th TFS. On April 28, 1967, he claimed a MiG-17 with two bursts of his 20mm gun, while flying F-105D 60-0504. The kill was confirmed by the USAF, but the North Vietnamese stated all MiG-17s returned safely that day. (USAF via Jack Gurner)

F-105Ds of the 354th TFS are ready at Takhli's flight line for the next combat mission. The canopy rail shows the name of Capt Charles Couch. He was one of the five F-105D pilots claiming a MiG-17 on May 13, 1967, which were all confirmed by the USAF. However, according to North Vietnam, not a single MiG-17 was lost that day ... (Jack Gurner)

Left: Capt Gene Eskew's name is on the canopy rail of F-105D 61-0127 *Snoopy* (354th TFS). The aircraft shows the MiG-17 kill marking Eskew claimed on April 19, 1967, as Panda 01, with 926 rounds of 20mm, as confirmed by the USAF. However, Capt Eskew told the authors he flew F-105D 62-4364 that day, which also received a MiG kill marking. The USAF confirmed four MiG-17 kills that day by F-105Ds, but the North Vietnamese stated that all aircraft RTB-ed (returned to base) safely. (Larry Hensley)

Below: A lineup of F-105Ds on last chance of the three Takhli Thud squadrons: 333rd TFS/RK, 354th TFS/RM and 357th TFS/RU. To the right, an EB-66 is taxiing to the runway. (via RAC/CAM)

Capt R. Patchett of the 333rd TFS is getting out of Squadron F-105D 59-1772 *Li'l Baby Lane* after flying his final combat sortie ("counter") on April 30, 1968. The welcoming team, a clean flight suit and champagne in a cooler are waiting for him. "772" sports two MiG-17 kill markings, one by Col Bob Scott, the Wing commander, on March 26 and one by Maj Harry Higgins, both of which were confirmed by the USAF. "There is a Way" refers to a 1967 USAF movie with the same title, showing there was a way for F-105 aircrews to complete 100 "counters" over North Vietnam, despite the heavy losses of men and aircraft. (via André Wilderdijk)

Above: In February 1969, 62-4394 is a F-105D of the 333rd TFS. Some of the M-117s have fuse extenders for a mission in Laos. It is named *Kilarney Count,* and the crew chief is SSgt A.S. Bowie. (Omar Wiseman)

Right: Although WW-F 63-8316 *The Nomads* is a 333rd TFS aircraft in April 1969, the aircrew – who are on last chance and showing their hands, meaning they are not touching their controls – are assigned to the 357th TFS. (Lt Col James Broussard)

Above: Four F-105Ds on the Takhli runway prior to take-off. All aircraft are configured with at least one ECM pod and six M-117s. (via RAC/CAM)

Left: F-105Ds 61-0188 (354th TFS) and 61-0220 (333rd TFS) during prestrike refueling. Both are configured with two AGM-12C Bullpups for use against targets in Laos. (USAF, John Evans)

Below: WW-F 62-4415 *Sam Seducer* (354th TFS) returns to Takhli after flying a combat mission over Laos. The F-105D is 62-4229 *Sneaky Coyote II* of the 357th TFS. (USAF, John Evans)

Above: 62-4284 (354th TFS) was the only triple-MiG killer F-105D, with two MiG-17s claimed by Capt Max Brestel on March 10, 1967, and one by Capt Gene Basel, both of the 354th, on October 27. Unfortunately, "284" and its pilot were lost in a crash on March 12, 1976, near McAllister, Oklahoma, while assigned to the 465th TFS at Tinker AFB. (USAF, A1C William Merritt)

Right: F-105D 60-0522 *Cheeze Maker Special* (357th TFS) is on last chance. The pilot's hands are visible to show maintenance personnel that they are off his controls. Two MiG kill markings are visible. (via RAC/CAM)

Below: WW-F 63-8323 *Whiskey Bill* (357th TFS) is taxiing to last chance in September 1968. (via Norm Taylor)

Above: The crew of WW-F 63-8313 *Road Runner II* (357th TFS) is letting down after flying a combat mission, probably an "easy" one to Laos as the aircraft is not configured with ECM pods. (USAF, John Evans)

Left: F-105D 62-4229 *Jeanie II* and *I Dream of Jeanie* (357th TFS) just prior to take-off from Takhli. The name on the canopy rail was that of Lt Col Jack Spillers, who, on July 8, 1970, became the Squadron's last commander at Takhli. (Jack Spillers)

Below: The pilot of F-105D 61-0161 (44th TFS) on finals to Takhli in December 1969, photographed by the USAF's SSgt John Evans in an accompanying WW-F.

Above: WW-F 63-8327 *Sweet Caroline* (44th TFS) in its revetment at its new home, Takhli. After the bombing halt on November 1, 1968, WW-Fs were primarily used as bombers on missions over Laos. While at Korat, the Squadron color was yellow. As the 357th's color was yellow as well, the "Vampires" adopted black. (Jim O'Neal)

Right: An avionics technician at work at Takhli on a 44th TFS F-105D. (Jim O'Neal)

Below: F-105D 62-4361 (44th TFS) taxis in on July 25, 1970, from Tainan where IRAN (Inspect and Repair when Necessary) was accomplished. It also received a fresh paint job. (Jim O'Neal)

Chapter 6

The 355th TFW and July 20, 1966

On July 1, 1966, the 355th and 388th TFW had a combined UE of 90 F-105D/Fs, 54 and 36, respectively. They had 130 assigned with 126 possessed, of which 55 were assigned to the 355th and 71 by the 388th TFW. Authorized were 115 crews, while 216 were formed, of which 211 were Combat Ready (CR). By July 31, the UE of the 388th was increased from 36 to 72 through the addition of the 13th TFS and 34th TFS, resulting in a combined authorization of 126 aircraft. The AOB for the 355th TFW showed 35 Ds and four Fs and 57 and five, respectively, for the 388th (101). Crew authorization was increased to a combined 189, of which 215 formed, with 213 being CR.

Losses

Operation *Rolling Thunder*, the aerial bombardment campaign against North Vietnam, was in full swing in July 1966, with the F-105 Thunderchief playing a "prominent" role. This was not only with regard to the number of combat sorties flown, but also as to lost and damaged aircraft. In July, 6,450 USAF attack sorties were scheduled and 6,170 flown, including 6,118 AR sorties. Including CAP and RESCAP sorties, the total flown was 7,124. Of the AR sorties, 4,916 were flown by Thai-based and 1,202 by South Vietnam-based aircraft. Of those 4,916, F-105D/Fs flew 3,310, plus all 52 strike sorties. In addition, 424 combat sorties were flown in Laos and 276 non-combat sorties, resulting in a total of 4,062 F-105D/F sorties. The 355th TFW's share was 1,754 scheduled with 1,709 flown, including 1,413 *Rolling Thunder* sorties.

Although the number of F-105 flying hours programed for July (based on the lower UE of 90 aircraft) was 7,315, 9,450 were flown, including 8,039 combat, 116 training and 1,295 "other" hours. This resulted in an average utilization rate of 72.5 percent, a daily sortie rate of 31.2 percent (which proved to be the highest for 1966), a 72.8 percent OR rate (72.6 percent worldwide) and a 23.7 percent NORM (Not Operationally Ready, Maintenance) rate.

Eighteen of the 21 USAF losses in July involved F-105D/Fs, of which 15 were on AR missions and three on *Iron Hand* (IH) missions, for a rate of .0054 percent (in May, the rate was even .0097 percent). Eleven of the F-105 losses were suffered in Route Packages (RP) 6A and 6B, the Hanoi–Haiphong area. The average number of 125.4 F-105s possessed at both bases meant that 14.35 percent of their aircraft were lost in combat. No F-105s were lost over Laos. However, there were also two operational losses, increasing losses to 15.95 percent.

As a matter of fact, in the April–July period, 48 F-105s were lost over North Vietnam, plus six operational losses. Three aircraft were lost over Laos, making a total of 57. In this respect, the authors of the July "Summary of Air Operations Southeast Asia" remarked that, if F-105 losses would continue as they had during that period, the entire SEA F-105 force would be lost and require replacement aircraft in about an eight-month period. And yet, in August, another 19 Thuds were lost, 18 over North Vietnam and one over Laos, while the number damaged was 15 and one, respectively. The 355th lost 15 aircraft. Five were lost on the 7th, "Black Sunday," two Fs and three Ds. Of the crew members, five were listed as MIA (missing in action) and two recovered. The 18 losses brought the cumulative number to 126. With 15 losses in Laos, the total was 141, which did not include operational losses. The number of damaged F-105s was 189 and 35, respectively (224).

"Firsts"

In addition to the assigned F-105D/Fs in the 333rd, 354th and 357th TFS, the Wing's aircraft roster also showed 12 Brown Cradle EB-66Bs and six EB-66Cs. These aircraft were flown by the 6460th TRS and 41st TRS, Photo Jet (PJ), respectively, which were assigned to the 460th Tactical Reconnaissance Wing at Tan Son Nhut, but attached to the 355th. This meant that all EB-66 maintenance was provided by the 355th. Tenants at Takhli were Det 1, 4258th Strategic Wing with 15 KC-135As and Det 2, 38th Aerospace Rescue and Recovery Squadron with three HH-43B Huskies.

According to its history, the 355th TFW and Takhli accomplished several "firsts" in July. On the 11th, Takhli became the first F-105 MOB in SEA, resulting in increased maintenance. To accomplish this, several facilities were built, including a Field Maintenance Shop and a maintenance hangar. The Wing, as the first F-105 wing, exceeded 10,000 combat sorties, reaching 10,024 on the 31st. Related to this, it also became the first F-105 wing to exceed 4,300 flying hours in one month (4,366), at the same time exceeding a monthly utilization rate of 75 hours/aircraft (76.7). Although the 388th TFW had received the first two of its six scheduled Wild Weasel III (WW III) F-105s by May 31, 1966, Takhli had to wait until July 4 before the first four of six Wild Weasel F-105Fs (WW-Fs) were welcomed. Nineteen days later, it lost its first F-105F, when 63-8338 of the 354th TFS was hit by a SA-2 in RP 6A, with both crew members listed as MIA (in August, three more WW-Fs were lost, while one F was severely damaged. As its final F was transferred to Korat, the 355th had zero F-105Fs assigned on August 31).

Next Phase

On July 9, the next phase in the *Rolling Thunder* interdiction campaign, 51 Alpha, was initiated. Its AR program was designated 51 Charlie. 51 Alpha included four new JCS-selected targets, which were split evenly between the USAF and US Navy. Of the 107 USAF strike sorties scheduled on July 9–31, only 40 were flown. Two of the USAF targets were the Ha Gia Highway Bridge on Route 3 (RP 6A), 19 miles north of Hanoi and the nearby (at a quarter-mile distance) Ha Gia Railroad Bridge (JCS target 18.23). Both targets were struck by F-105Ds on July 20. Eight aircraft struck the highway bridge, while JCS 18.23 was struck by four of the 16 sorties flown that day by the 333rd TFS "Lancers." The four Thunderchiefs formed Stinger flight, expending eight M-118s. BDA indicated interdiction of the approaches of and possible minor damage to the highway bridge. By July 31, it was serviceable again. As to the railroad bridge, BDA showed its rail approaches were cut.

In addition, 98 other sorties were flown by Thunderchiefs for a total of 110 of 206 USAF sorties flown that day. However, not all went according to the screenplay, as Takhli lost two aircraft that day, one F-105D and one EB-66C. The F-105D loss involved 61-0116, flown by the Wing's Vice Commander, Col Bill Nelson, who was attached to the 354th TFS for flying. He was Shotgun 01 in a flight of four Ds on an AR mission in RP 6A. While on his second run strafing trucks, "116" was hit by an 85mm shell. His flight searched visually and by radio for 15 minutes with negative results. SAR efforts were not attempted due to lack of information and area defenses. It was assumed Nelson went in with his aircraft. His initial MIA status was later changed to KIA.

To support the strike forces in RP 6A with ECM and ELINT (electronic intelligence), EB-66Cs Devil 01 and 22 were paired, each with six crew members. Devil 01, 56-0464, had been spare and filled in for the original aircraft that aborted due to engine problems. The aircraft were MIGCAP-ed by two F-4C Phantoms and two F-104C Starfighters. Devil 01 was hit in the right fuselage at 29,000ft by an SA-2, northwest of Thai Nguyen. Devil 22's aircrew reported the incident and observed two parachutes just prior to aircraft disintegration at approximately 5,000ft, but no beepers were heard or parachutes seen. Initially, all six crew members were listed as MIA. It became clear later all had ejected and five were captured, being released in 1973. The remaining crew member was declared KIA on January 18, 1978.

Capability

It was standard 7AF policy to frag the 355th TFW with 62 sorties a day against a UE of 54 F-105. This had been within the Wing's capability during the first six months of 1966. However, subsequent to June 30, several factors developed, requiring the number of F-105s available to be increased above 54 if the Wing was to continue to meet the 62 sorties per day. Through August 6, this resulted in a series of messages to higher headquarters, with the first message dated July 14.

After the Wing assumed MOB responsibilities, a message, "Number of F-105 Aircraft Authorized 355 TFW," was sent to 7 and 13AF et al on the 14th, reiterating its request to increase its UE. In the end, 63 F-105D/Fs would result in enough airframes to meet 7AF's mission requirements, while performing all required maintenance. The 63 UE would amount to 54 aircraft CR and nine in various stages of maintenance at all times. The Wing also pointed out a second major problem: a severe overcrowding of available aircraft parking space at Takhli. Although the Wing requested a 63 UE, it stated that until such time as the number of KC-135s was reduced to ten on station, it would not be possible to accommodate over 58 F-105s, including those undergoing maintenance.

Referring to its July 14 and 19 messages (explaining a reduced number of available F-105s), the 355th, on July 20, sent a message "Reduced F-105 Capability 355 TFW" to 13 and 7AF. It was explained that the Wing was unable that day to fill all sorties as fragged by 7AF due to the large number of aircraft out of commission for heavy maintenance and recovery of F-105s at locations other than Takhli. In addition, the Wing repeated that a UE of 54 was inadequate to meet the fragged missions, perform MOB major phase inspections and support of base depot-level Time Compliance Technical Orders (TCTOs). At 05:00L, 43 Ds and five Fs (48) were possessed at Takhli. In addition, two possessed aircraft were not available, with 59-1732 out of commission at Da Nang and 62-4343 with battle damage at Udorn. Four Ds were assigned but were possessed by Air Force Logistics Command (AFLC) and being repaired by RAM Teams; 62-4261 as of June 11, 61-0080 as of June 23, 60-0454 as of June 30, and 62-4239 as of July 9. All but "080" received battle damage. Returning from a combat sortie, the pilot of "080" experienced an AC generator failure and landed at Udorn. After touchdown, the drag chute failed to deploy, and the aircraft travelled the full length of the runway, continued off the end, shearing the nose wheel. The pilot deplaned uninjured. The aircraft was shipped to CONUS on September 29. Of the possessed aircraft at Takhli, 36 were OR, 32 Ds and four Fs. The 12 aircraft out of commission were so due to several reasons. For example, two Ds were undergoing major phase inspections and two an engine change, while TCTO 1F-105D-687 (the addition of AGM-45 capability) was being accomplished on one D.

The Wing was fragged for 30 sorties in the morning, while 28 were launched due to one abort and one Maintenance Non-Delivery. As two aircraft were used to spare Stinger and Fosdick flights, they were not available to cover the two losses. Thirty-two sorties were on the afternoon schedule, with the first take-off at 12:00L and the last at 13:45L. Upon recovery of the AM mission, the 12:30L aircraft status showed that the number of available aircraft had dropped to 46: 41 Ds and five Fs, as two Ds had recovered at Udorn. Only 21 Ds and three Fs were OR and available for the PM mission, as 14 Thuds of the AM missions were out of commission and not available. Twenty-two (20 Ds and two Fs) were ultimately launched. One F-105D aborted and was replaced with an F-105F in Locust flight (*Iron Hand*). This meant that no aircraft were available to fill the remaining ten sorties, resulting in the cancelation of Elm and Bobcat flights and the reduction of the Vernon flight to two aircraft. The Wing concluded its message that its reduced operational capability would continue to exist until the number of in-commission aircraft was increased to 54, 49 Ds and five Fs.

Although the UE was not immediately increased, on July 22, PACAF and, on July 27, 5AF informed the Wing that the number of assigned F-105s would be increased to 62, effective August 15. In an August 6 message, "F-105 Aircraft Assignment," to 5, 7 and 13AF, the 355th once more repeated why the number of assigned F-105D/Fs had to be increased, (1) the assumption of the F-105 MOB status; (2) the reduction of the periodic inspection interval on the J-75 engine from 200 to 125 hours. Due to the shortage of J-75s, this factor resulted in a further reduction of three to four aircraft a day; and (3) the sharply increased battle damage and loss rate in July, with 11 losses and 19 aircraft damaged. As it took three to four days to replace a lost aircraft, the reduction quantitatively varied from two to five aircraft. The combined effect resulted in a net reduction of eight to 12 F-105s available for operational use. The Wing stated that 49 Ds and four Fs were possessed and on station (53), of which 44 Ds and four Fs were OR (48). The Wing concluded its message, requesting the three Air Forces for all action possible to attain 62 F-105D/Fs on station at the earliest possible date.

After the Wing lost an F-105 in combat on August 4, with two receiving major battle damage, the Wing requested the 6441st TFW at Yokota three replacement aircraft. The 6441st, by August 5 message, informed the 355th that further replacements were not forthcoming. In an August 12 message, "F-105 Aircraft Support," to 13AF, 7AF referred to 355th TFW's August 4 message, stating 7AF was concerned over the status of F-105 support after the recent increase in loss and battle damage rates. Sorties were being canceled daily due to the non-availability of aircraft. Replacement aircraft were needed on a timely basis in order that units could meet combat commitments. In order to maintain the Wing at authorized strength, 13AF was requested to implement procedures to immediately deploy F-105 replacement aircraft to SEA upon notification or loss. The Wing's AOB on December 11 showed 55 Ds and 8 Fs (63). However, on the final day of December the numbers had dropped to 47 and seven respectively (54). The UE was ultimately increased to 72 F-105s, but only after the 44th TFS had relocated from Korat to Takhli and reassigned to the 355th on October 15, 1969.

Thunderchiefs were still in the "open" at Takhli in early 1966, before revetments were built. At least five of the aircraft are still uncamouflaged. One tail still shows the 335th TFS markings and another those of the 562nd TFS. (Capt Stanley Gunnerson)

In June 1966, most, if not all, F-105Ds had been modified with RHAW equipment, even the still-uncamouflaged F-105D. In the background, SAC's KC-135As. Col Bob Scott tried to get the tankers off his base, but to no avail. At least not until March 1968, when they moved to U-Tapao RTAFB. (USAF)

Although stationed at Takhli and attached to the 355th TFW, the two EB-66B/C units were assigned to the 460th TRW at Tan Son Nhut. On May 17, 1966, USAF approved the redesignation of RB-66Cs to EB-66Cs and of B-66Bs to EB-66Bs. EB-66C 54-0459 (41st TRS, Photo Jet [PJ]), is taxiing past F-105s in maintenance. (USAF)

An EB-66C of the 41st TRS, PJ, taking off from Takhli. (USAF)

Right: The first (daylight) pathfinder or "buddy" bombing mission was flown on January 1, 1966, in Laos by two Brown Cradle B-66Bs and 12 F-105Ds. Its K-5 navigation/bombing system provided precision bombing information to, for example, F-105Ds pilots, enabling them to bomb from high altitudes, regardless of weather conditions. The photo shows a B-66B as pathfinder for four F-105Ds. (USAF)

Below: A flight of four 355th TFW F-105Ds. Only the uncamouflaged F-105D, 59-1731, still showing 562nd TFS markings, has RHAW equipment applied. (USAF)

F-105D 62-4405 (354th TFS) is configured with two M-118 3,000lb bombs on the inboards. Tail codes were only applied later in 1967. Col Bob Scott was adamantly against putting big white letters on the tails of his aircraft, fearing it might make discovery by North Vietnamese and North Korean MiG pilots easier. He lost his case. (USAF)

63-8341 was one of the two WW-Fs arriving at Takhli on July 20, 1966. The other one was 63-8338, which was lost three days later on the mission where Buddie Reinbold's F-105D was damaged. 341's service life was not "boring" either. On August 16, 1966, it received battle damage. After expending 643 man-hours, it was flown on September 29 to Tainan for further repair and completed on January 7, 1967, after another 5,047 man-hours. However, on April 19, the aircraft was bagged by the cannon fire of a North Vietnamese MiG-17. (Lou Chesley)

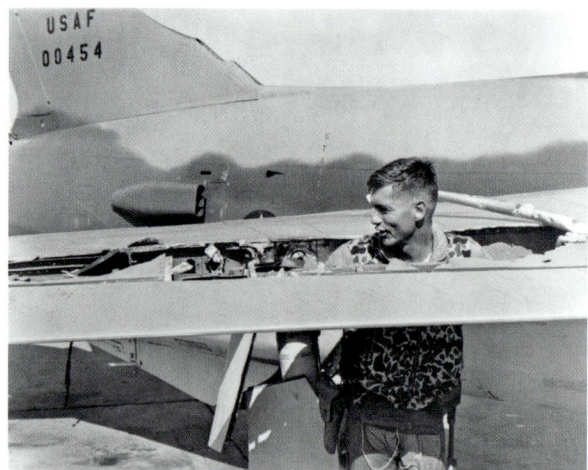

Maj William McClelland was on TDY from Yokota to the 333rd TFS and overseeing the damage on his aircraft, 60-0454, inflicted by a 85mm round while striking a bridge spanning the Kuih Bich Dong River, on June 30, 1966. The round hit the right inboard pylon and exploded. All structure was torn out for approximately 4ft. Despite a large amount of drag, he managed to fly "454" 500 miles to Takhli, where he landed without incident after a successful in-flight refueling. Repair was completed on October 4, requiring 1,898 man-hours, meaning "454" was not available on July 20. (USAF)

Above left: F-105D 59-1748 (333rd TFS) was another aircraft possessed, but not available on July 20. It was damaged on July 14 and flown to Udorn on the 24th for repair by a RAM Team. (via RAC/CAM)

Above right: Also unavailable on July 20 was F-105D 62-4261. It is seen at Da Nang, where its pilot landed "clean" after being hit by AAA on June 11. Repair was completed on August 6, at the expense of 1,813 man-hours. (Coll/TvG)

The one WW-F aircrew that was recovered on August 7 consisted of Capts Ed Larson (pilot) and Mike Gilroy (EWO), of the 354th TFS. They were Mambo 01 in 63-8358 on a suppression mission in RP 6A. The target was a railroad marshalling yard on the Northeast Railroad. Their armament included two AGM-45s on the outboards, two LAU-3/A pods on the inboards, and the gun. Prestrike refueling was over the Gulf of Tonkin and ingress north of Haiphong harbor. After Gilroy had called a SAM site dead ahead of them, Larson armed a Shrike and expended it a few seconds later against its Fan Song radar. Approximately 15 seconds after launch, the radar's signal went off the air, probably meaning it was hit. After a second SAM site started to track "358," Larson lined it up with the new threat and expended his second Shrike. In the meantime, it became clear by a shrill howl over the intercom that the site had launched at the aircraft. The first SAM exploded harmlessly, but a second one roared out of the clouds and exploded just in front of "358." It was extensively damaged, but still controllable, and Larson directed it towards the Gulf of Tonkin. There, they safely ejected and were recovered by the crew of a HU-16B Albatross of Det 1, 3rd Air Rescue and Recovery Group at Da Nang. (USAF)

Chapter 7
Southeast Asia Farewell to the F-105 Thunderchief

In the first months of 1969, an Air Staff study, "Reduction of Forces in SEAsia," addressed four options if a reduction in forces was to be directed. Option C, for instance, would deploy the F-105D/F fleet from Thailand to Kadena. On March 16, 7AF reacted to the study in a message to PACAF. It stated: (1) that much of the F-105's survivability in the North Vietnamese air defense environment was due to the continual development of the ECM posture; (2) the loss of this capability would materially reduce its ability to be responsive to directives to resume the bombing; and (3) the Wild Weasel/*Iron Hand*/Combat Nail capability of the F-105 squadrons would be sorely missed, and losses should increase accordingly. The defensive environment in Laos was already intensely hostile. Should these aircraft be removed, the protection afforded the B-52Ds used in Operation *Arc Light* missions would be lost. Should radar-controlled guns be introduced along the Ho Chi Minh Trail in number, the F-105s would also be required to perform the *Iron Hand* role in this area. 7AF's conclusion was that Option B (inactivation of two F-100D/F squadrons in South Vietnam and delaying or cancelling their replacement by A-37s) was the least undesirable, as it degraded its capability to a lesser degree.

Keystone

After his meeting with South Vietnam's president Nguyen Van Thieu on June 9, 1969, President Richard Nixon announced a redeployment from South Vietnam of 25,000 personnel by August 31, with six future incremental withdrawals to be examined in the light of several criteria, including progress in training and equipping of South Vietnamese forces. The seven *Keystone* redeployment increments involved USAF forces in South Vietnam only. As *Keystone* redeployments began, US military leaders deemed it imperative that force levels in Thailand not be decreased and hopefully even be increased. They must have been disappointed, as, on September 30, Nixon announced the withdrawal of 6,000 personnel from Thailand by June 30, 1970, and of 9,865 personnel the following year. Mounting budgetary restrictions were also debit to the Thai redeployment. The latter included the PCS of the 44th TFS to Takhli to maximize management efficiency, making the 355th TFW a four-squadron F-105D/F wing; the former included the inactivation of the 355th, the redeployment of 12 WW-Fs to Korat and the remainder to Kadena and CONUS, and the closure of Takhli. The redeployments were dubbed *Banner Sun*.

Rationale

On February 1, 1970, 13AF issued its "Bed-down and Deployments Program," which contained projected force changes in Thailand, but for planning purposes only. As for the 355th TFW, it stated its inactivation with the 44th and 333rd TFS, and the PCS of the 354th and 357th TFS to Kadena (all four with 18 UE) in July. Takhli was to be closed in the first quarter of 1971. In his February 19 memorandum, "Force Planning – Thailand," SECDEF Melvin Laird requested the JCS to provide a plan by April 1, with two redeployment packages of about 5,000 personnel each, to reduce the US forces in Thailand to 32,200 by June 30, 1971.

USAF Program Directive (PD) 72-1 (cut-off date was January 5, 1970, and publication date February 20) documents reflected withdrawal of all F-105 forces from SEA in September 1970: the 44th TFS would PCS to Kadena, the 333rd TFS inactivate, the 354th TFS would PCS/WOPE to Misawa (Japan), and the 357th TFS inactivate. The 67th TFS would PCS/WOPE to Kadena from Misawa. The 18th TFW would then have three F-105 squadrons assigned, the 12th with six F-105D/Fs and 18 F-105Gs, the 44th and the 67th TFS each with 24 F-105D/Fs. As PD 72-1 reflected withdrawal of all F-105 forces from SEA, the JCS, in a March 25 message "F-105 WW Program," requested CSAF and PACAF to provide options and rationale to continue WW-F support capability in SEA. PACAF reacted two days later by stating that rationale was forwarded to CINCPAC on March 18, 1970. According to PACAF, it was logical, in the face of a continually declining force posture in SEA, to retain the most flexible and versatile firepower available, the F-4. It was recognized that the F-105 redeployment would eliminate WW and *Iron Hand* support for B-52 missions and reduce USAF's ability to provide surface-to-air missiles (SAM)/AAA suppression support, but the increased F-105 force at Kadena would provide a rapid reentry capability when required. It was also stated that alternatives were considered, particularly with regard to retaining a WW capability at Korat, but PACAF did not recommend the retention of an F-105 squadron in lieu of the F-4.

Banner Sun

On April 30, 1970, JCS sent its JCSM-202-70 memorandum to Laird with redeployment plans of selected US forces from Thailand. The plans were approved by Laird in a June 5 memo and included the Package A deployment of the 333rd and 354th TFS to CONUS and the 44th and 357th TFS to Kadena, all to take place in September, with Takhli closing in December. Some 3,170 personnel were involved. However, Laird also stated a requirement to maintain a WW capability to reduce the risk to B-52D *Arc Light* forces. CINCPAC was authorized by JCS on August 21 to proceed with *Banner Sun*. It was also indicated that the F-105 redeployment to Kadena was subject to modification in view of a SECDEF requirement for JCS to review a USAF West Pacific deployment posture. USAF responded one week later, and the posture included, among other things, plans to deploy all F-105s from Thailand to CONUS, except for those aircraft required for one 24 UE F-105 WW squadron at Kadena, providing the capability to maintain a detachment in Thailand with six aircraft to reduce risk to *Arc Light* forces. On September 12, CINCPAC directed execution of *Banner Sun*. As JCS had not yet determined the F-105 redeployment destinations, they would be the subject of separate action. An early decision was requested, as USAF had run out of time in terms of accomplishing the necessary operational planning and scheduling: the original stand-down scheduled for September 1 had already slipped to the 15th, with another slippage being faced. When CINCPAC was asked by JCS/J-3 Operations on September 15 to confirm CINCPAC's concurrence as reflected in previous messages, CINCPAC replied on the 16th in a message "Redeployment of F-105s from Thailand" that the proposal to provide a squadron with WW capability at Kadena was concurred with, but not with the loss of two F-105 squadrons from PACOM (Pacific Command), as it would represent a significant reduction in tactical air force capabilities. However, two days later, JCS authorized the redeployment of four F-105 squadrons to CONUS and the retention of a 24 UE F-105 WW squadron at Kadena, providing the capability to maintain a six-aircraft detachment in Thailand. Authorization was given to proceed with the CONUS redeployments.

Guidance

In the meantime, the last new pilots from CONUS had arrived at Takhli in June 1970. On the 30th, the 355th TFW was authorized 72 F-105D/Gs and possessed 63: 51 Ds and 12 Gs. On July 1, the Wing published OPlan 309-70, providing broad guidance and direction to effect the relocation of all F-105s. Referenced was the Wing's PAD 70-4 "Base Closure Plan, Takhli RTAFB, Thailand" of March 31, 1970.

Appendix I: Annex A "Relocation Schedule" in Change #2 of October 1:

Number	Destination	Date
6 F-105Gs	Korat	September 24
5 F-105Ds	Kadena	October 10
18 F-105Ds	McConnell	October 19
18 F-105Ds	McConnell	October 26
Remainder	Kadena/Korat*	October 25–28

* not firm

Appendix II: Annex A "Forces" in Change #2 was based on CSAF messages of August 10 and September 29, and 13AF's October 1 message:

Unit	UE	Action	Disposition of aircraft*
355TFW	–	Inactivate	
44TFS	18	Inactivate	CONUS
333TFS	18	Inactivate	CONUS
354TFS	18	PCS/WOPE to Misawa	CONUS
357TFS	18	Inactivate	CONUS

* 23rd TFW at McConnell. Nineteen aircraft possessed by the 12th TFS were also to transfer to the 23rd. The 11 F-105Gs possessed by the Takhli squadrons were to transfer to the 12th TFS, of which six were to go to the Forward Operating Location (FOL) of the 12th, Det 1, to be established at Korat. A twelfth F-105G, 62-4446, was in a non-flyable condition.

Coronet Bighorn

On September 14, 1970, 13AF directed the cessation of tactical F-105 missions the next day, except for WW operations, while authority was given to implement OPlan 309-70 with an in-place date for the 190 personnel and six aircraft on the 25th. With regard to the former, the 13AF directive was rebutted immediately by 7AF, as SAC tanker support for the CONUS redeployment would not be available until mid-October. As a result, 13AF informed the 355th that its tactical missions would continue until 15 days prior to the F-105 redeployment. The Wing informed 13AF on October 1 that 51 F-105Ds were available, including ten for transfer to the 18th TFW and 11 that were committed for modification to the Thunderstick II configuration.

The Wing's final combat mission was flown on October 6, when four F-105Ds struck a "relatively minor target" in Laos. The next day, the stand-down began, with flying terminated to allow for transfer inspection, washing, painting and other routine maintenance. On the 8th, PACAF established November 24 as the Wing's inactivation date. The next day, the 355th was informed that PACAF had received approval to increase the number of WW aircraft at Korat from six to 12 through May 1971.

"Coronet Bighorn, TAC/USAF strike frag order to TAC/AFSTRIKE OPlan 100" was the message on October 10, identifying the deployment of individual squadrons as follows, each with 18 aircraft, in three cells of six, with 30-minute intervals:

Bighorn	Squadron	From	Date
I*	333rd	Takhli	October 19
II**	357th	Takhli	October 26
III**	44th	Kadena	November 2

  * October 19–23. The force commander was Lt Col Ron Frazier. One D from the second cell aborted into Korat with an AC generator failure and returned after repair. An F-105D of the third cell lost all instruments and returned to Takhli, a spare taking its place. The route was across Thailand, Laos and Da Nang. East of Da Nang, flights would rendezvous with KC-135s, and then proceed across the Philippines to Andersen. There, F-105D 60-0488 had the aft section removed by the deployed EST, to repair a ruptured hydraulic line, but it was completed in time. After 36 hours on the ground, they departed for Hickam, crossing over Wake Island and just south of Midway. Another 36 hours were spent at Hickam. All 17 arrived on October 23 at McConnell, where the pilots were given a warm welcome by 23rd TFW personnel.

** Due to typhoon activity in the South China Sea, remaining aircraft departed on the 27th, 18 to CONUS and five to Kadena. Due to maintenance problems, only 17 aircraft departed for CONUS, where they arrived on October 30. The five Kadena aircraft launched about an hour after the CONUS group. They were originally scheduled to be part of the Kadena–CONUS deployment on November 2. As the aircraft had non-repairable wing cracks, no wing tanks could be carried, and the five were not cleared for the flight across the Pacific. They were cleared for a one-time flight to Kadena, configured with the 650-gallon centerline tank. At Kadena, the aircraft would be assigned to the 18th TFW. The lineup included Majs Ken Mason and Russ Bartlett, Capt Ron Brekke and 1Lts Dave Sawyer and Chuck de Vlaming. Three made it to Clark, stayed overnight and flew the next day to Kadena through Tainan, where they were refueled. Brekke and De Vlaming had to return to Takhli, as their 650s would not take fuel from the tanker. On the 28th, they and 1Lt Lanoux, who had diverted into Korat on the 27th, joined up with an F-4 to lead them to Clark. However, the Phantom's crew developed fuel problems and returned to Da Nang, accompanied by the three F-105Ds, as none of the pilots were flight lead-qualified. The next day, the aircraft departed and proceeded to Clark. The three F-105D pilots eventually made it to Kadena. However, Ron Brekke ended up spending six days at Tainan, waiting for his aircraft to be repaired. He was the last Takhli pilot to make it to his assigned destination. Two of the pilots joined *Bighorn III* with 12 aircraft, Kadena–CONUS, on November 2. The pilot of F-105D 61-0050 aborted into Wake Island with engine vibration problems on the 4th, resulting in an engine change.

Retiring the Flag

However, there was still one Thunderchief left at Takhli, F-105F 62-4446, which was damaged in an accident in August 1970. The aircraft, assigned to the 354th TFS but flown by a 44th TFS crew, was extensively damaged on the 19th when it landed excessively hard and ran off the runway after returning from a night mission. It was regarded as a major accident. On September 17, 1970, SMAMA requested the aircraft be prepared and air shipped to McClellan for repair/modification. After dismantlement, packing and crating, "446" was loaded aboard a C-133 Cargomaster and flown on November 10 to SMAMA. "446" was to be one of the two final Thunderchiefs to leave TAC, the 35th TFW and George on October 8, 1980. On December 2–5, three cargo aircraft airlifted the sixth and final cargo package to Byrd Field to assist the Virginia Air National Guard to convert from F-84Fs to F-105D/Fs. The airlift shipments took 44 aircraft, most C-141s, to transport 699 short tons of equipment.

    Col Clarence Andersen, the Wing commander, retired the flag on October 12, marking the end of SEA combat flying. A 12-aircraft flyover was performed to commemorate the event. Almost two months later, on December 10, the 355th TFW with the 357th TFS and maintenance and support squadrons were inactivated through PACAF Order G-259, December 8. On the same date, the 44th and 354th TFS were reassigned to 13AF WOPE. PACAF MO-32, December 14, PCS-ed the 333rd TFS WOPE to McConnell with a Transfer Effective Date of October 15. It was then reassigned to the 23rd TFW as a non-operational unit.

Left: On October 12, 1970, Col Clarence Andersen, the 355th TFW commander, retired the flag, marking the end of the Wing's combat flying in SEA. A 12-aircraft flyover was performed to commemorate the event. (USAF)

Below left: Another picture of the flyover. (Coll/TvG)

Below right: Although the Romeo Mike tail code of F-105D 62-4387 was removed, the blue-colored radar reflector denotes it was assigned to the 354th TFS. (Maj McDaniel)

Right: Two CONUS-bound Thuds with their tanker over the Pacific. (Dean Vikan)

Below: This "bread truck" was decorated appropriately for the arrival of the 355th TFW Thunderchiefs at McConnell. Some 48 aircraft were put into temporary storage before reassignment to Air National Guard and Air Force Reserve units. (USAF)

WW-F 62-4446 was the final F-105 to leave the 355th TFW and Takhli. It was damaged in an August 1970 accident. As it was not repaired in time, it was dismantled, packed and crated, loaded aboard a C-133 Cargomaster and flown on November 10 to SMAMA for repair. "446" was nicknamed *The Silent Majority* with the names *Pat* and *Linda* below the cockpits. (Dean Vikan)

Korat-1

After 13AF stated on August 6, 1970, that "recent information indicated a six UE F-105G Wild Weasel FOL be established at Korat when USAF activities ceased at Takhli," the 18th TFW sent a team to Thailand on September 3 to survey requirements of a Wing detachment at Korat and to coordinate transfer of equipment to Kadena and Korat.

PACAF informed USAF on September 4 that the phasedown of USAF activities at Takhli required the relocation within Thailand of six F-105 WW aircraft. As this was not considered to occur under the *Banner Sun* movement constraints, authorization was requested to relocate six aircraft from Takhli to Korat as soon as possible. Ten days later, 13AF informed the 355th and 388th TFWs that plans were approved for the relocation by September 25. In addition, 13AF asked 7AF to reduce the frag rate to accommodate it. On the 15th, 13AF directed the relocation of 22 officers, 168 airmen and six F-105Gs with a September 25 in-place date.

Det 1, 12th TFS, was activated at Korat on the 18th, per PACAF Order G-209, September 16, 1970, and attached to the 388th TFW. Its basic mission was to provide support for *Arc Light* B-52D missions, while the secondary mission was to act as a hunter/killer or search/destroy weapons system against any SAM or radar-controlled AAA sites that posed a threat to friendly forces. Five days later, a nine-person advance party arrived at Korat, including its commander, Lt Col Clarence Sonderman, and the assistant Ops Officer, Maj Sam Hauck. On the 24th, the F-105 War Reserve Spare Kit arrived. Five F-105Gs departed for Korat on the 24th after a flyover ceremony at Takhli, with the sixth arriving one day later. The Det also assumed operational commitments that day and the first combat mission was flown. The 354th TFS, for instance, contributed three of the nine required aircrews and two of the six F-105Gs. The move also involved 123 AGM-45 Shrikes and 128 AGM-78 Standard ARMs. In a September 26 message, PACAF recommended to CSAF that Det 1, 12th TFS, be inactivated and replaced by the 44th TFS, coinciding with the move of the F-105s to CONUS. Assignment would be to the 388th TFW. When the requirement for a WW capability in Thailand would be terminated, the Squadron would relocate WOPE to Kadena and be reassigned to the 18th TFW. HQ USAF was asked to issue a movement directive. The request, however, was disapproved.

After it was determined the number of F-105Gs was not adequate to support the continued high frag rate of *Arc Light* B-52D strikes, 7AF requested on October 3 to move six additional aircraft, crews and support personnel to Korat. Approval was received on the 9th, with the final aircraft arriving on October 19. In an October 29 message, the 388th TFW informed PACAF that 321 of the 354 authorized personnel and the 12 aircraft were on station. Eight of the aircraft were received from 12th TFS assets and four from SMAMA. As the required additional aircraft and support personnel necessitated the formation of a squadron, Det 1, 12th TFS, was inactivated on November 1, 1970, by PACAF G-245 of October 30. The Det suffered no losses.

Korat-2

G-245 also activated the 6010th Wild Weasel Squadron (WWS) and assigned it to the 388th TFW. It absorbed mission, personnel and equipment of Det 1. Up to now, WW-Fs had frequently been called F-105Gs, which was technically incorrect, as only WW-Fs that had been configured with ALQ-105 ECM pods were to be designated as F-105Gs. Yet, the 6010th WWS history for the first quarter of 1971 stated that the "F-105 fully equipped with the APR-35/36/37 and the AGM-78B package has been redesignated the 'G' model." The Squadron now had F-105Fs and Gs assigned.

Although the UE remained at 12, the number of aircraft was increased in March, April and May and used to sustain a temporary surge in operational commitments. For this reason, three F-105Gs were received in March from the 23rd TFW and two from the 66th Fighter Weapons Squadron (FWS) at Nellis AFB. On June 30, 1971, the 6010th still possessed two aircraft with the older receiver equipment

(APR-25/26 and ER-142). F-105 63-8306 was the "high-time" F-105 in the second quarter (April to June) with 114 hours and 18 minutes, while 63-8320 had most sorties, 44 in April. On July 1, all aircraft had been modified with the APR-35/36/37, while four aircraft were also equipped and operational with the ALQ-105 jamming system. Due to repair, Korat's runway was closed for a certain number of hours each day. During this period, September 16–November 27, an FOL was established at Udorn. About one-third of the F-105F/G sorties were flown from the FOL for 256 launches and recoveries. On September 21, the Wing launched eight F-105F/Gs, 34 F-4Es and eight EB-66C/Es in support of Operation *Prize Bull*; a massive attack on key targets in the southern part of North Vietnam. To reduce the number of four-digit unit designators, PACAF G-273 of November 16 inactivated the 6010th WWS, effective December 1. The Squadron suffered three losses.

Korat-3

G-273 also activated the 17th WWS on December 1, which also assigned it to the 388th TFW, replacing the 6010th. The mission essentially remained the same. Due to the loss of several aircraft to SAMs and AAA in the latter part of 1971, Operation *Proud Deep Alpha* was executed December 26–30 against targets in Route Pack 1, with Thunderchiefs flying 55 sorties. On December 31, the number of authorized aircrews was 18, of which 17 were formed and available. The aircraft UE was 12 with two Non-Operating Aircraft (NOA). Fifteen F-105F/Gs were assigned, of which 14 were possessed, with ten being combat-ready. There were still a few airframes without the ALR-31 SAM warning device, but most aircraft had the ALQ-105.

As of January 4, 1972, the 17th WWS was called on to fly WW support for AC-130s that were "working" the trails and enemy concentrations in the *Steel Tiger* area in Laos. Missions were all flown at night, with a Weasel on-station above the Spectre continually from 17:00–05:00L. In a February 19 message, 7AF and 13AF informed PACAF that an agreement was reached that the F-105 UE for the 388th TFW would be increased to 15, plus two NOA for fragging purposes. However, on March 25, the UE was still 12 F-105Gs, with 19 assigned for the B-52D *Arc Light* surge.

After the large-scale March 29 invasion of South Vietnam by the North Vietnamese, supported by Viet Cong forces, JCS directed Tactical Air to be augmented by forces from other PACAF bases and CONUS. As to the latter, McConnell's 561st TFS was directed on April 6 to deploy 12 F-105Gs to Korat to augment the WW force. Deployment was made under a "Bare Base" concept as *Constant Guard I* and also included 36 F-4Es from Seymour Johnson to Ubon and six EB-66s from Shaw AFB to Korat. As the Squadron flag remained at McConnell, the deployed personnel and aircraft became known as Det A(lpha), 561st TFS. The aircraft departed on April 7–9, a cell of six F-105s on each day. On the 8th, the Det was attached to the 388th TFW. Seventeen C-141 Starlifters airlifted 373 personnel and 430,000lbs of cargo. Closure was April 16. The TDY was initially for 90 days, but this was changed to 179 days. After arrival of Det A, the 17th WWS training section checked out the aircrews as rapidly as possible, so they could relieve pressure on the 17th. As all 561st crews had served in SEA before, theater indoctrination was reduced to a minimum, but yet were briefed on local area procedures, applicable WW tactics, and rules of engagement. Det A flew its first combat mission on April 12 and had its own training program initiated within two weeks.

In an April 8, 1972, message, 13AF informed the 388th TFW that the 17th WWS would be assigned 21 aircraft to support increased combat sortie requirements, including 17 UE and four NOA. The additional aircraft arrived in April. Large-scale strikes on North Vietnam were initiated on April 8 with the *Freedom Train* series of operations, followed on May 12 by Operation *Linebacker*. June saw the replacement of the Zulu Bravo code on the tails of the 17th WWS to Juliet Bravo, which was more in line with those on Wing aircraft. It appears the tail code on Det A's Thuds was changed around July 25 from Motel Delta to Whiskey Whiskey. After the 388th Organizational Maintenance Squadron was activated on August 1, 1972, the 17th WWS lost its maintenance personnel, with only 39 officers and nine airmen authorized.

On September 22, a "new" WW aircraft arrived at Korat, when the 67th TFS deployed six WW F-4C IVC Phantoms from Kadena under *Commando Flip*. The deployment became necessary after TAC's proposal was approved to return six of its F-105s and nine aircrews to CONUS to establish a WW training force for SEA to replace TDY personnel. It was agreed that six of the present seven F-105Fs would be returned, which still lacked the ALQ-105 modification. The six aircraft departed Korat on August 25 in two cells of three as *Coronet West 11* and *13*. Ninety-three support personnel, a spare engine and 104,000lbs of equipment were also redeployed. The WW F-4Cs returned to Kadena on February 15, 1973. Of the 1,588 sorties flown by the 17th WWS in the 3rd quarter, 1,205 (combat support) involved the support of more than 2,000 B-52 cells (of three aircraft each). No BUFFs were lost due to enemy action. The last mission in *Linebacker I* was flown on October 27. WW-F/Gs flew 520 combat sorties, losing three aircraft. To prepare for *Linebacker II*, the Wing "stood down" on December 17. The all-out bombing effort was initiated the next day. F-105Gs flew 98 support sorties as 24-hour protection during B-52D/G sorties in RP 6A and 6B. In addition, F-105Gs flew 64 daytime TACAIR (Tactical Air) support sorties. WW F-4Cs sorties numbered 40 and 62, respectively. In January 1973, the 17th flew 34 *Arc Light* support sorties in central and southern North Vietnam, with the final one on the 15th, and in northern South Vietnam. On the 27th, all offensive operations against North Vietnam were terminated. Three days later, WW activity over Laos resumed with B-52 support sorties in the *Steel Tiger* and SAM suppression in the *Barrel Roll* areas of Laos. These sorties, however, were few in number and were terminated on February 21. B-52 support sorties were initiated over Cambodia on March 10. In addition, F-105Gs started escorting F-111As with tactics being similar to *Arc Light* escort missions. Both Det A (the 27th) and the 17th WWS flew their last combat sorties in April, and, on May 1, both entered a total training environment.

It then became necessary to create a training mission, which kept the aircrews proficient in B-52 support tactics. The mission was planned, briefed, led, and flown as a combat support mission. A variety of enemy threats were simulated and both offensive and defensive tactics employed. The Sky Spot radar site (part of Operation *Combat Sky Spot*) at Nakhon Phanom RTAFB (NKP) was utilized for RHAW training, while simulated dive bomb and road reconnaissance missions were initiated. On June 1, ECM/EW training, utilizing US Navy (USN) ships in the Gulf of Tonkin, was proposed by 7AF. The training, nicknamed Weaselex, was initiated on August 12 by employing F-105Gs/EB-66Es against Navy fire control radars. To get to the "target" area, aircrews were allowed to overfly Laos, Cambodia and South Vietnam. All missions were fragged as F-105G four-ships and involved aerial refueling. Through September 30, 95 sorties were flown. On September 5, the 17th had flown its first training sortie at the newly improved Chandy bombing/gunnery range in Thailand.

The Squadron's total training environment allowed PACAF, on June 13, 1973, to approve the Wing's May 10 proposal to redeploy the Det A aircrews to CONUS (by mid-June, only one Det A aircrew was still at Korat). The 12 aircraft and 180 maintenance personnel, however, were to be retained to support the Wing's aircraft training program. However, on August 25, PACAF directed the Wing to have the 12 F-105Gs ready for redeployment to George on/about September 5. Two days later, TAC sent a warning order for the redeployment of one F-105G and two F-4E squadrons, and 12 F-111As to commence on September 5. The Thunderchiefs were then prepared, ensuring all were in the best possible condition. Deployment, as *Coronet Bolo I,* was initiated at 03:30L on September 6, with three cells of four F-105Gs. The aircraft were flown by 561st TFS aircrews, who arrived at Korat on August 31. The route was Korat–Andersen–Hickam–George with an EST at Andersen and Hickam, while Wake Island was designated as the emergency base for missed air refueling and aborting aircraft. One aircraft aborted on take-off from Korat due to afterburner problems, which necessitated a single fighter-tanker deployment, launching on the 7th as *Coronet Bolo IA*. The two aircraft trailed the others by 24 hours. The Andersen–Hickam leg on September 7 also did not proceed as advertised, as three aircraft air-aborted. The first one returned to Andersen, being unable to obtain fuel during the first refueling, the second had hydraulic failure and recovered at Wake, while the third F-105G had to divert to Johnston Island due to severe engine vibrations, resulting in an engine change. The stragglers arrived at George at 16:00L on September 14. Equipment was airlifted by C-141s, while four C-130s airlifted 154 passengers and three tons of excess baggage. With regard to the Det's return to CONUS, Adm Noel Gaylor, CINCPAC, stated on September 7, "You are great SAM killers. Happy to have had you on the team. Nice going."

In September 1973, the 17th WWS was authorized 12 F-105Gs, and had an average of 13 assigned and 12 possessed. The number of aircrews was 27 authorized and 26 formed and available. Weaselex training sorties were canceled on October 14 due to the necessity for F-105Gs to overfly North Vietnamese AAA sites in South Vietnam. However, new routes were developed, and training was resumed on the 29th. For a variety of reasons, all training, including Weaselex, sorties were canceled on November 10 for the remainder of the quarter. One reason was the fuel crisis and a second the possibility of Russian trawlers collecting ELINT information on USN weapon systems. Of the 1,260 flying hours allocated for the fourth quarter, the 17th only utilized 593 in 420 sorties. For instance, in November, only 105 sorties were flown for 93 hours and 24 minutes. Training was not halted altogether, however, as on November 18 and December 16 two Korat aircrews were sent TDY to Nellis to attend a continuation training course at the mid-point of their tours. This was regarded to be of vital importance. In a December 13 message, TAC informed PACAF that the 57th Fighter Weapons Wing (FWW) at Nellis had the capability to accommodate continuation training during two weeks for one PACAF F-105G aircrew through February 1974.

However, the problem of WW training continued into the new year, with no suitable sites available in Thailand and insufficient slots at Nellis to provide training for all aircrews. In the first quarter of 1974, F-105G aircrews participated in six *Commando Scrimmage* exercises, which fully tested various aspects of the 388th TFW's combat capability and emphasized coordination between the several dissimilar aircraft within the Wing. F-105Gs flew as hunter-killers and simulated enemy interceptors. Only two exercises were held in the second quarter. In the August 29–September 16 period, the 17th WWS flew 21 ELINT sorties around the NKP area in a (fruitless) search for threat signals from AAA/SAM activity. Flights consisted of at least two aircraft per week.

After a September 29 JCS execute message, 13AF issued PAD 75-13-5, providing guidance to the 388th TFW and 17th WWS for the relocation of the Squadron to CONUS, as *Coronet Exxon*. R-day was established as October 30, and George was the destination. The early relocation was approved due to the lack of adequate training facilities and meant the end of a little over ten-year period of combat operations by the Thunderchief. Involved were 211 personnel, 187 tons of cargo and 12 F-105Gs (the 13th, 63-8304, was at Sacramento Air Logistics Center). On October 8, PACAF published Aircraft Assignment Directive 74-10-073, reassigning the aircraft to TAC. On the same date, TAC published Frag Order 74-13 "Coronet Exxon", which listed the 35th TFW as the receiving wing. Change #1 was issued for the change in departure date and on the 15th, the Squadron stood down.

Although departure was scheduled on October 30, it was brought forward by one day in order to not disrupt the King of Thailand's visit to the Korat area. The route was Korat–Andersen–Hickam–George, with ESTs at Andersen and Hickam. Each cell was to be accompanied by five U-Tapao KC-135s. The first four aircraft departed at 06:19Z on October 29, followed by two cells at 20 minute intervals. However, 62-4446 aborted for a battery high charge light, resulting in a 27-minute late take-off (its crew, Lt Col Gordy Walcott, Squadron commander, and Capt Walter Kennedy, was flight lead and "claimed" 63-8316 in another cell) and 62-4442 due to a utility hydraulic failure. Both aircraft departed at 12:50Z on the 30th. This meant that "446" was not only the final Thunderchief to depart Takhli, but also the last one to depart Korat and PACAF. Call signs were Width 11-14, 21-24, 31-32 and 33-34. After arrival at George on November 1 and 3 (the stragglers), the aircraft were assigned to the 562nd TFS, which was activated on October 31 and assigned to the 35th TFW. Personnel and cargo departed in October and November, with the last personnel leaving on December 15. With the departure of the F-105Gs, the 17th WWS became non-operational. PACAF GA-46 of November 7 inactivated the Squadron effective November 15.

Epilogue

To give an idea about the number of sorties, we look at the strike, armed recce and flak support sorties flown in the July 1, 1965–October 31, 1969, period and compare those of the F-105D/F and F-4C/D/E.

	F-105D/F	F-4C/D/E	Total*	Percentage of F-105D/F	Percentage of F-4C/D/E
North Vietnam	71,845	70,361	151,268	47.5	46.5
Laos	51,576	84,665	225,039	22.9	37.6
Total	123,421	155,026	376,307	32.8	41.2

* Includes all aircraft types

In the Air War, the USAF suffered the loss of 1,676 fixed-wing aircraft n combat. For the F-105, this number was 334, of which 282 were in North Vietnam (30 by SAMs and 22 by MiGs), 51 in Laos and one in South Vietnam: four undisclosed, 293 Ds, 26 Fs, and 11 Gs. The loss rate was 2.1:1,000 sorties. This was 3.3 in North Vietnam, meaning roughly that one out of every four F-105s hit in combat would crash. For comparison, 358 F-4C/D/Es were lost for a 0.7 loss rate. The overall probability of crew member survival in the F-105 given a loss was 65 percent. For South Vietnam, this was 100 percent, for Laos 56.9 percent, and North Vietnam 66.2 percent. There were also 63 operational losses, 53 in Thailand.

With regard to these numbers, a couple of things should be noted. For example, (1) aircrews faced concentrated and modern AAA, including radar-guided 85mm and 100mm guns, which was by far the biggest culprit; (2) especially in the early days of the Air War, the US had no answer to the SA-2, supplied and supported by the Soviet Union; (3) Rules of Engagement (RoE) were dictated by "Washington"; and (4) the bases of the Soviet Union-supplied MiG-17, -19, and -21 aircraft were off limits until JCS issued *Rolling Thunder* 55 Execute Order on April 22, 1967, to attack Hoa Lac (USAF) and Kep (USN) airfields.

Aircrews	Cambodia	Laos	North Vietnam	South Vietnam	Total
Rescued	0	32	99	1	132
Captured	0	1	99	0	100
Missing	0	14	96	0	110
Killed	0	8	18	0	26
Total	0	55	312	1	368

For instance, the "Bulldogs" of the 354th TFS lost 59 F-105D/Fs. Thirty-nine of their crew members were listed as either MIA or KIA, while 28 were recovered. Ninety-two crew members, of which three were attached for flying, completed 100 "counters."

A F-105G of Det A, 561st TFS being prepared for the return flight to George. (USAF)

WW-G 62-4416 of Det A on the brink of returning to CONUS, George AFB, on *Coronet Bolo*. The yellow pod on the outboard is a luggage pod. (USAF)

The three red stars shown on F-105G 63-8320 denote that its aircrews downed three North Vietnamese MiGs. However, none was officially confirmed by the USAF. According to legend, in 1967, combat circumstances forced the aircrew to drop all pylons and ordnance, which hit a MiG-17 and resulted in a crash. The second kill, on December 19, 1967, was attributed to Majs William Dalton and James Graham. The kill itself was confirmed by the USAF, but F-105F 63-8329 was the aircraft involved. A typo perhaps? The third kill was also claimed in December, but never confirmed. As Tom Brewer of the Air Force Museum wrote, "I feel fairly confident that 320 DID NOT score." (USAF)

Pilot and EWO being assisted by the crew chiefs to get strapped in prior to taking off for the first leg (to Andersen) of the flight to George. (USAF)

A flight of four Det A aircraft take off from Korat for the last time. (USAF)

In late October 1974, it was also time for personnel and equipment of the 17th WWS to return to CONUS, George, on *Coronet Exxon*. WW-G 62-4425 is being configured with a 450-gallon fuel tank on the right inboard. (USAF)

Left: The "17th WWS Thunderchief Team," personnel who performed the various tasks to allow the aircraft's crew to accomplish their mission. (USAF)

Below: When the WW-Gs of the 17th WWS were returned to CONUS, their aircrews seemed not to have the luxury of a luggage pod. (USAF)

Above: WW-G 62-4442 over the Pacific while in the air refueling process with a KC-135. (USAF)

Right: One of the F-105 pilots shot down, becoming a "guest" of the North Vietnamese and later being released, was Maj Wayne Waddell. He arrived at Takhli in April 1967 and was assigned to the 354th TFS. On July 5, 1967, he was one of the three 355th TFW pilots who were downed over North Vietnam, while on his 47th "counter." Maj Waddell was #4 in Wolf flight, flying F-105D 61-0127. The target was the Cao Nung Railroad Yard, JCS 18.24, some 40 miles northeast of Hanoi. While releasing his ordnance, the aircraft was hit by 85mm/100mm AAA. The other flight members saw Maj Waddell coming off the target at 6,000ft while "127" was burning and saw it crash. No chute was seen or beeper heard and Wayne was listed as MIA, as were the other two pilots. He was captured almost immediately by a small group of civilians with one or two rifles. Wayne recalled: "I ejected at very low altitude and my parachute only opened enough to break my fall before I was sitting on the ground. Attempts to talk to anyone on my survival radio were futile and I considered I would be reported as KIA. I was sent to Hoa Lo prison, 'Hanoi Hilton', in Hanoi." Several days later, Wayne was told to put on his flight suit again and someone else's boots. With arms tied behind him and blindfolded, he was taken to a truck. After driving around for some time, they stopped, and Wayne was taken out. After the blindfold was taken off, he saw he was on a dike in a rice paddy. A young North Vietnamese woman appeared beside him, armed with a rifle with a bayonet attached. He also noted two male Caucasians with camera/film equipment. Wayne was directed to walk along the dike toward the camera/film crew with his head down and to not talk. This was repeated several times, after which he was trucked back to Hanoi. He heard later the crew was from East Germany. In one segment of the film *Pilots in Pajamas*, the narrator stated that "Downed air pirate Major Dewey Waddell was now securely behind lock and key in Hanoi." Wayne was released on March 4, 1973. (via Wayne Waddell)

Other books you might like:

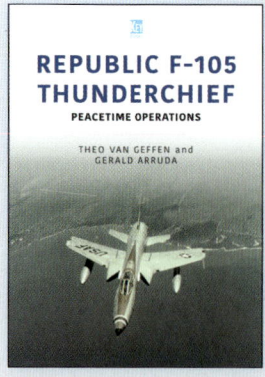
Historic Military Aircraft Series, Vol. 6

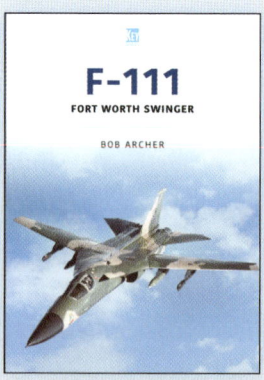
Historic Military Aircraft Series, Vol. 3

Historic Military Aircraft Series, Vol. 9

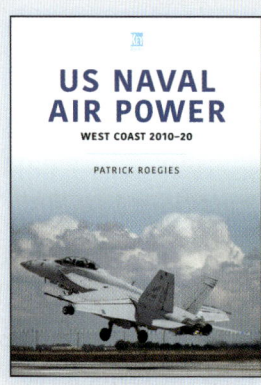
Air Forces Series, Vol. 2

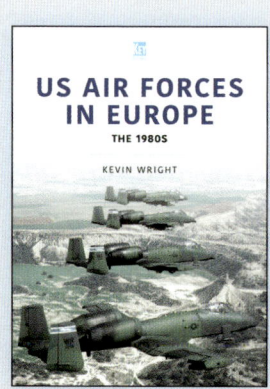
Air Forces Series, Vol. 4

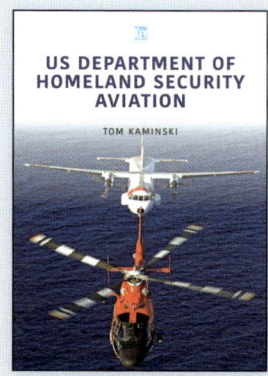

For our full range of titles please visit:
shop.keypublishing.com/books

VIP Book Club

Sign up today and receive TWO FREE E-BOOKS

Be the first to find out about our forthcoming book releases and receive exclusive offers.

Register now at **keypublishing.com/vip-book-club**

Our VIP Book Club is a 100% spam-free zone, and we will never share your email with anyone else. You can read our full privacy policy at: privacy.keypublishing.com